Math Motivators

Grades 4–6

Grades 4–6

Math Motivators

PUZZLES, GAMES, BULLETIN BOARDS, AND SPECIAL MOTIVATORS

Marian R. Bartch
Jerry J. Mallett

Scott, Foresman and Company
Glenview, Illinois London

ISBN 0-673-18265-7

Printed in the United States of America.
3 4 5 6—MAL—90 89 88

About This Book

Mathematical drill and practice exercises are often described as "dull," "boring," or "the same old thing" by both students and teachers. Such exercises are, however, essential for mastering the basic facts of every arithmetical operation. The function of this book is to provide teachers with a wide variety of activities and games to spark and maintain students' interest in skills-work practice. The material is divided into five sections for easy selection of appropriate activities.

I Games for Individual Instruction
II Games for Group Instruction
III Puzzle Sheets
IV Learning Boards
V Special Motivators

Each game or activity is arranged in mathematical sequence—from the basic foundation skills to more advanced operations. Each focuses on a specific skill and is keyed to an approximate grade level. Detailed directions for construction and use are provided to ensure quick and easy adaptation for any classroom. Moreover, the accompanying patterns of game components can be copied and used immediately. Also included is a skills listing, which classifies each game or activity under a specific skills heading.

Most of the puzzle pages are self-correcting and can be duplicated just as they are. The learning boards may be easily reproduced by using an opaque projector. All materials needed for the games and activities are readily available, making this book an accessible and convenient teaching aid.

By using the materials in this book as supplementary exercises in the classroom mathematics program, teachers will keep their students' interest and motivation high as well as enable them to master important and necessary mathematical skills. The wide range of skills and grade-level suitability permit a more individualized approach for teachers who strive to meet the unique needs and abilities of each student.

The authors wish to thank Carol George and Anne Cernak for their helpful reviews.

Contents

Mathematical Skills

PART I
Games for Individual Instruction

Mathematics learning requires an emphasis upon drill and practice exercises that reinforce or maintain the initial instruction. These exercises, though necessarily repetitious, need not be "boring" if they are varied through the use of games and other creative activities. Concept understanding and skills-mastery vary greatly among students in any given classroom. Individualization of instruction and pacing of learning can be achieved by using self-checking games. The games' built-in self-checks allow students to receive immediate feedback and strengthen their grasp of mathematical principles. This feedback also tells teachers whether more practice is needed in certain concept areas.

The individualized, self-check games in this section are easy to make, for construction details are clearly described. The materials used in most of the games—posterboard, color markers, scissors, glue—are readily available to the classroom teacher. Other materials are inexpensive and easy to collect.

The games are arranged in mathematical sequence from the easiest skills to the more difficult.

Runaway Stagecoach

Skill Reinforced: Multidigit addends/Regrouping tens

Materials Needed:
- 1 sheet of $8\frac{1}{2}'' \times 11''$ white posterboard
- 10 sheets of $2'' \times 2\frac{1}{2}''$ green posterboard
- scissors
- felt-tipped pens of various colors

Construction Directions:
1. Mark the large posterboard piece using an enlargement of the gameboard pattern shown here.

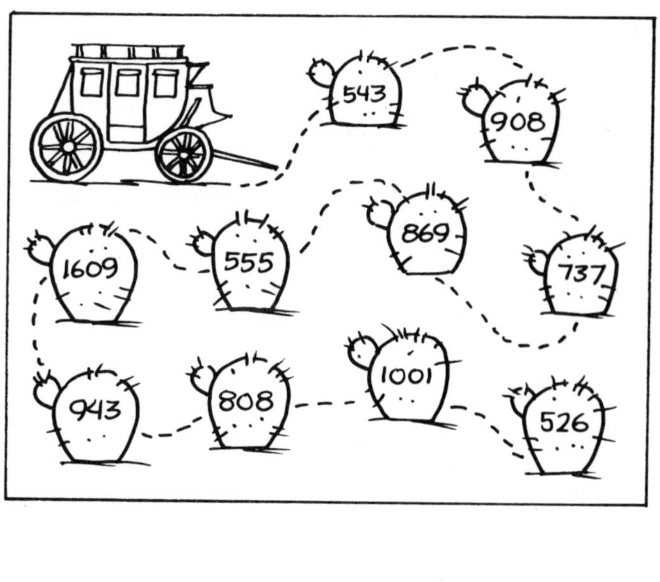

2. Cut and mark the green posterboard pieces using the cactus pattern shown below.

3. Copy each of the following problems onto a different cactus as shown in the illustration. Write the corresponding answer on the back.

182	426	565	488	283
+361	+482	+172	+381	+272
543	908	737	869	555
829	661	512	411	235
+180	+282	+296	+590	+291
1009	943	808	1001	526

Game Play:
1. Place the gameboard on your desk. Spread out the cactus plants, problem side up, next to the gameboard.
2. Now follow the runaway stagecoach until you come to a cactus. On it you will find an answer to one of the cactus-plant problems. You are to find the right cactus and place it over the cactus on the gameboard. Continue until you reach the stagecoach.
3. When you are finished, turn over all the cactus plants for the self-check.

Thunder and Lightning

Skill Reinforced: Multidigit addends/Regrouping ones and tens

Materials Needed:
- 10 sheets of 4″ × 7″ gray posterboard
- scissors
- felt-tipped pen

Construction Directions:
1. Cut, fold, and mark the posterboard pieces as shown above.

2. Copy the following problems and answers onto the gameboards in the manner shown in the illustration.

254	396
+258	+147
522, 512	543, 553
454	568
+276	+352
720, 730	920, 910
637	725
+273	+189
810, 910	914, 924
827	926
+184	+295
1011, 911	1121, 1221
364	489
+186	+762
550, 560	1251, 1261

3. Print the word "Correct" under the flap with the right answer.

Game Play:
1. Choose a gameboard and look at the problem. Which of the "lightning answers" is correct?
2. When you think you know, lift up that flap for the self-check.

Horace Horseshoe

Skill Reinforced: Single-digit subtrahend and double-digit minuend/Regrouping

Materials Needed:
- 12 sheets of 5″ × 8″ gray posterboard
- scissors
- felt-tipped pen

Construction Directions:
1. Cut and fold the posterboard pieces using the pattern shown here.

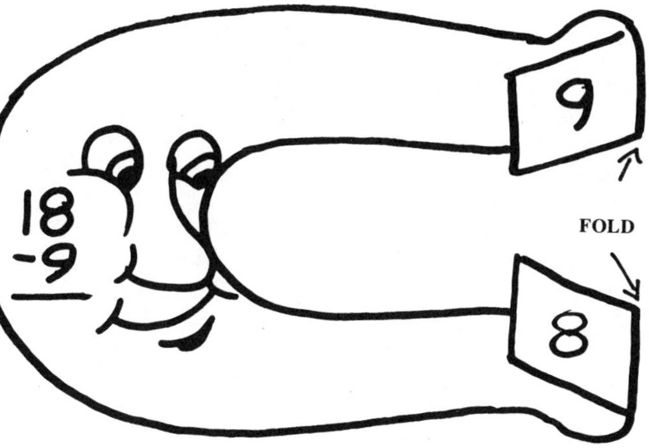

2. Copy each of the problem-and-answer sets onto a different horseshoe as shown in the illustration. Mark a smile face under the flap that has the correct answer.

Problem	Answers	
18 − 9	9	8
22 − 5	16	17
33 − 5	27	28
44 − 8	36	37
57 − 9	48	47
65 − 9	57	56
76 − 7	69	68
81 − 3	78	77
92 − 8	83	84
31 − 4	27	26
25 − 9	16	17
42 − 9	32	33

Game Play:
1. Take up one of the horseshoes, making sure the flaps are closed.
2. Look at the problem. Do you know the answer? The correct answer is on one of the two flaps.
3. When you think you know which answer is correct, turn up the flap to see if you are right. If you are right, you will see a smile face.
4. Do the same with all the horseshoes.

Tick, Tack, Subtract

Skill Reinforced: Double-digit subtrahend and double-digit minuend/Regrouping

Materials Needed:
- 1 sheet of 6″ × 6″ yellow posterboard
- 1 sheet of 3″ × 6″ white posterboard
- 9 sheets of 2″ × 2″ red posterboard
- felt-tipped pen

Construction Directions:
1. Mark the yellow posterboard piece as shown here.

8	19	14
39	26	42
37	16	7

2. Copy each of the following problems onto a different piece of red posterboard. Do not copy the answers on these pieces. Copy both the problems and their answers onto the white piece of posterboard for the self-check.

Problem	Answer
25 − 18	7
27 − 19	8
36 − 17	19
42 − 28	14
57 − 18	39
65 − 39	26
71 − 29	42
83 − 46	37
94 − 78	16

Game Play:

1. Place the large yellow gameboard and all the small red cards face down on your desk. Now look at the large gameboard. This game is like tic-tac-toe. You must cover three squares in a row in order to win. They may be in any of these directions:

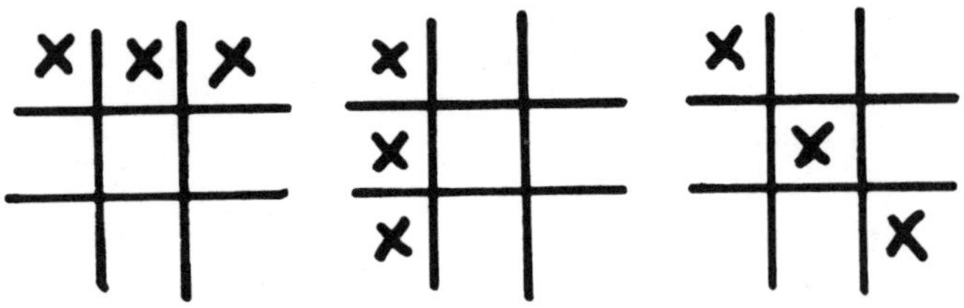

2. Turn over any five of the red cards. These are the only five you may use. Each card contains a problem. You are to find the answer to the problem on the gameboard and place the card over the answer. Do this with all five of the cards.
3. Did you cover three squares in a row and win? If not, turn the five cards back and flip five new ones over. Continue the game until you win. Use the self-check white card to see if you are correct.

Wise Old Owl

Skill Reinforced: Double-digit multiplicand and one-digit multiplier/No regrouping/Identify element

Materials Needed:

- 9 sheets of 7″ × 7″ white posterboard
- felt-tipped pens in various colors
- scissors

Construction Directions:

1. Cut and mark the posterboard pieces using an enlargement of the owl pattern shown above.

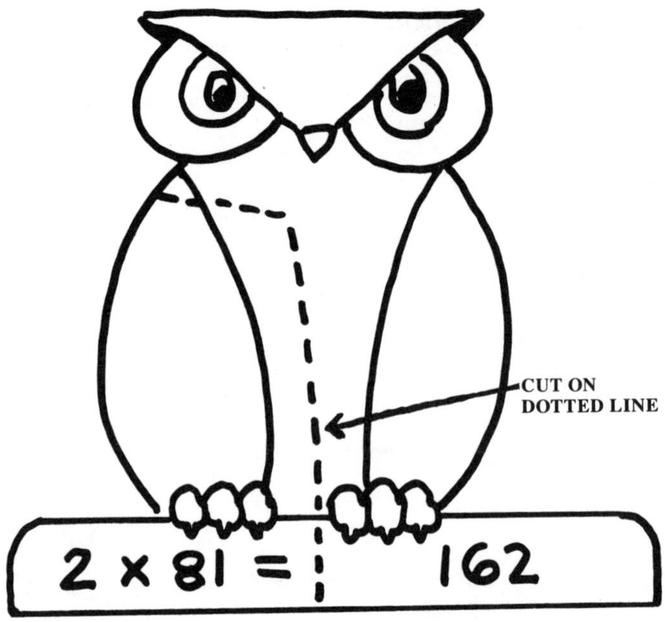

2. Copy each of the following problems and answers onto different owls as shown in the illustration. Cut all the owls apart, using a different pattern for each, similar to the one shown in the illustration.

$1 \times 91 = 91$	$2 \times 81 = 162$
$3 \times 72 = 216$	$4 \times 61 = 244$
$5 \times 51 = 255$	$6 \times 41 = 246$
$7 \times 31 = 217$	$8 \times 21 = 168$
$9 \times 11 = 99$	

Game Play:

1. Place the pieces with the problems in a pile and spread out the other pieces on your desk.
2. Look at one of the problem pieces. Do you think you know the answer? It is on one of the pieces that is spread out on your desk.
3. When you think you have found it, place it next to the problem piece. If it makes one complete owl, you are correct.
4. Do the same with the rest of the pieces. See how many owls you can put together.

The Devil Made Me Do It

Skill Reinforced: Three-digit multiplicand and single-digit multiplier/Multiple regrouping

Materials Needed:
- 6 sheets of 5″ × 7″ orange posterboard
- 6 sheets of 3″ × 9″ gray posterboard
- scissors
- felt-tipped pen

Construction Directions:
1. Cut and mark the orange posterboard pieces using an enlargement of the devil pattern shown here.

2. Copy each of the following problems onto a different devil as shown in the illustration.

2 × 885 =	5 × 329 =
3 × 548 =	8 × 222 =
4 × 426 =	9 × 185 =

3. Cut the gray posterboard pieces using an enlargement of the pitchfork pattern shown here.

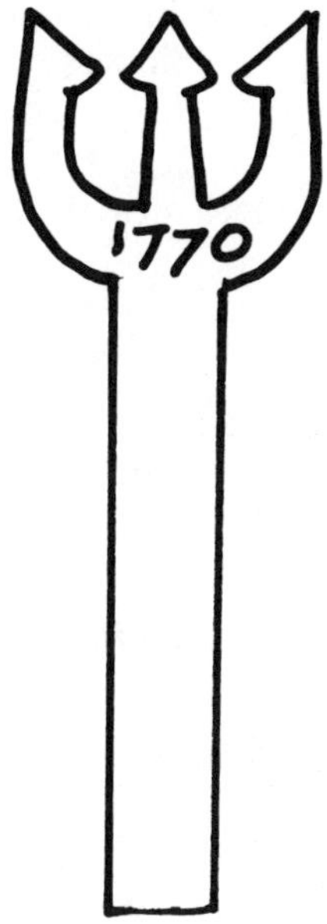

4. Write the following answers on the pitchforks as shown in the illustration.

1770, 1644, 1704, 1645, 1776, 1665

5. Write the correct answers on the backs of the devils for the self-check.

Game Play:
1. Place the devils in one pile and the pitchforks in another.
2. Take up one of the devils and look at the problem. Do you know the answer? It is on one of the pitchforks; see if you can find it.
3. When you think you have the answer, place that pitchfork next to the devil. Continue doing the same with the rest of the game pieces.
4. When you are finished, turn over the devils to see whether you have matched the pieces correctly.

Gilbert the Ghost

Skill Reinforced: Single-digit divisor and double-digit dividend/Regrouping/Remainders

Materials Needed:
- 1 sheet of 5″ × 8″ white posterboard
- scissors
- felt-tipped pen

Construction Directions:
1. Cut and mark the posterboard piece using the ghost pattern shown below. Then cut the piece apart on the dotted lines.

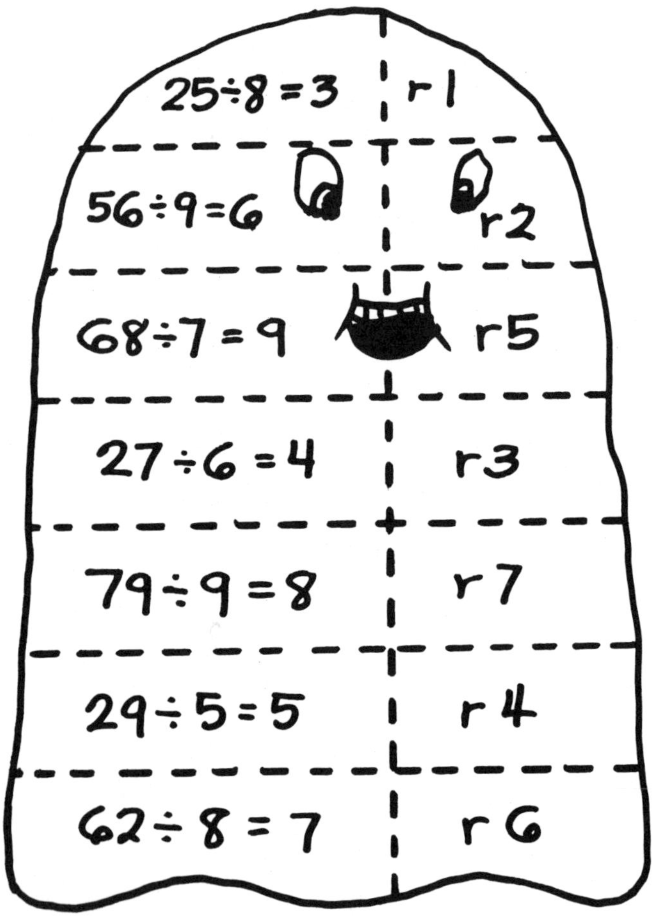

Game Play:
1. Place the problem puzzle pieces in one pile and the remainder pieces in another.
2. Take up one of the problem pieces and consider the "answer" given. You will see that a remainder is missing. Look through the remainder pieces and see if you can find the right one.
3. When you think you have it, place it next to the problem piece. Continue doing the same with the rest of the game pieces.
4. When you have finished, see if you can arrange all of them to form Gilbert the Ghost.

Sign in the Sky

Skill Reinforced: Multidigit dividend and single-digit divisor/Remainders

Materials Needed:
- 10 sheets of 2″ × 6″ white posterboard
- 10 sheets of 3″ × 7″ posterboard in various colors
- 10 shoelaces
- paper hole punch
- scissors
- tape
- felt-tipped pen

Construction Directions:
1. Cut and mark the 3″ × 7″ posterboard pieces using the airplane pattern shown here.

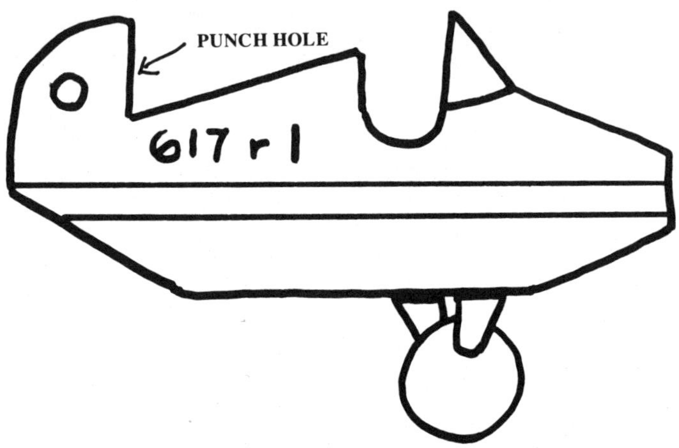

2. Attach a shoelace to each of the 2″ × 6″ posterboard pieces by placing it through the hole, as shown in the illustration, and taping it to the back.

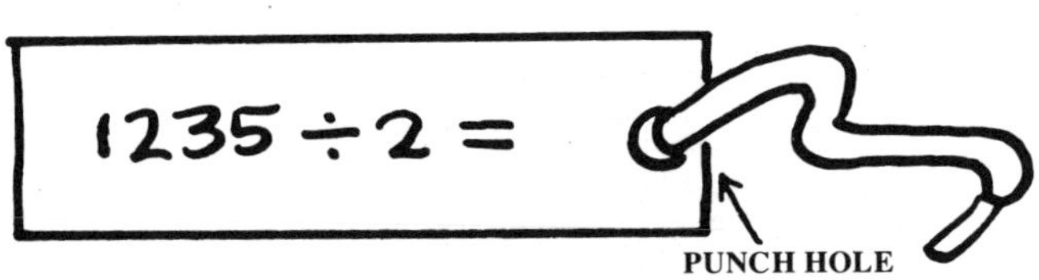

3. Copy each of the following answers onto different airplanes and each of the problems onto different sky signs, as shown in the illustrations. Write the answers again on the backs of the corresponding sky signs.

1235 ÷ 2 = 617 r1	3688 ÷ 3 = 1229 r1
4599 ÷ 4 = 1149 r3	6682 ÷ 5 = 1336 r2
5344 ÷ 6 = 890 r4	9847 ÷ 7 = 1406 r5
2957 ÷ 8 = 369 r5	7958 ÷ 9 = 884 r2
5947 ÷ 2 = 2973 r1	9982 ÷ 3 = 3327 r1

Game Play:

1. Place the airplanes in one pile and the sky signs in another.
2. Take one of the signs and look at the problem on it. Do you know the answer? It is on one of the airplanes. Look for it.
3. When you find it, attach the airplane to the sky sign by placing the shoelace in the hole in the airplane. Continue doing the same with the rest of the game pieces.
4. When you are finished, turn over the sky signs to see if you are correct.

Pumping for Oil

Skill Reinforced: Multidigit subtrahend and multidigit minuend/Multiple regrouping

Materials Needed:

- 10 sheets of 6″ × 9″ white posterboard
- 10 sheets of 3″ × 3″ white posterboard
- 10 sheets of 1″ × 7″ white posterboard
- 20 paper fasteners
- scissors
- felt-tipped pens in various colors

Construction Directions:

1. Mark, cut, and attach all the posterboard pieces together using the pattern shown below.

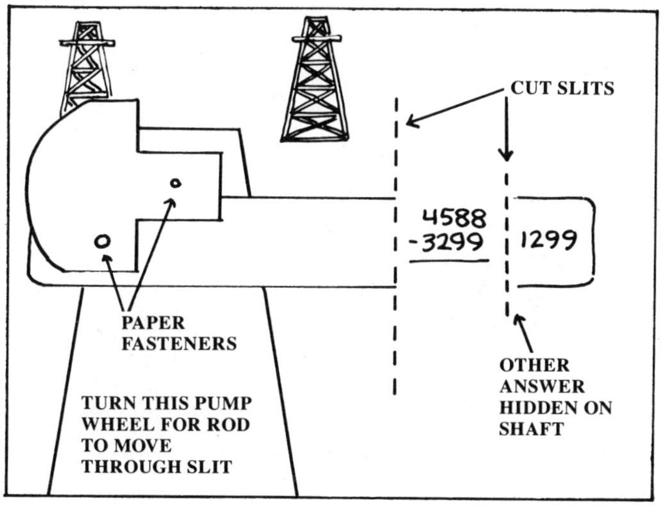

2. Copy each of the following problems and answers onto a different gameboard in the manner shown in the illustration. Simply turn the pump wheel to write in the two possible answers per problem. Copy the correct answer on the back of each card.

$$\begin{array}{r} 4588 \\ -3299 \\ \hline 1289,\ 1299 \end{array} \qquad \begin{array}{r} 8976 \\ -5597 \\ \hline 3379,\ 3279 \end{array}$$

$$\begin{array}{r} 7605 \\ -1276 \\ \hline 6329,\ 6429 \end{array} \qquad \begin{array}{r} 3656 \\ -1478 \\ \hline 2178,\ 2168 \end{array}$$

$$\begin{array}{r} 5611 \\ -2399 \\ \hline 3212,\ 3312 \end{array} \qquad \begin{array}{r} 6252 \\ -5188 \\ \hline 1064,\ 1054 \end{array}$$

$$\begin{array}{r} 9232 \\ -8145 \\ \hline 1087,\ 1187 \end{array} \qquad \begin{array}{r} 2767 \\ -1479 \\ \hline 1288,\ 1188 \end{array}$$

$$\begin{array}{r} 9654 \\ -2479 \\ \hline 7175,\ 7075 \end{array} \qquad \begin{array}{r} 5323 \\ -4136 \\ \hline 1187,\ 1177 \end{array}$$

Game Play:

1. Look at the problem on one of the gameboards. Do you know the answer?
2. Turn the pump wheel and look for the correct answer.
3. When you think you have found it, turn over the gameboard for the self-check.

Tulip Teasers

Skill Reinforced: Double-digit multiplicand and double-digit multiplier/Regrouping

Materials Needed:
- 12 sheets of 6″ × 6″ tan posterboard
- 12 sheets of 4″ × 7″ posterboard in various colors
- scissors
- felt-tipped pen
- color markers or stars

Construction Directions:

1. Cut and mark the tan posterboard pieces using the flower pot pattern shown here.

2. Cut the 4″ × 7″ posterboard pieces using the tulip pattern shown here.

3. Copy the following problems onto the tulips and the answers onto the pots. Affix same-color markers or stars to the backs of the corresponding pieces.

27	19	25	36	22	44
×12	×18	×17	×13	×19	×14
324	342	425	468	418	616

39	59	55	68	48	62
×17	×12	×16	×13	×19	×15
663	708	880	884	912	930

Game Play:

1. Place the tulips in one pile and the pots in another.
2. Look at the problem on one of the tulips. Do you know the answer? It is on one of the pots. Look for it.
3. When you find it, place the tulip stem into the opening of the pot and continue doing the same with the rest of the game pieces.
4. When you are finished, turn over the game pieces to see if you are correct. Each pair should have the same-color markers.

Goofy Grapes

Skill Reinforced: Three-digit multiplicand and double-digit multiplier/Multiple regrouping

Materials Needed:

- 3 sheets of 8″ × 8″ purple posterboard
- 15 purple posterboard circles with 1″ diameters
- scissors
- felt-tipped pen

Construction Directions:

1. Cut and mark the large posterboard pieces using the bunch-of-grapes pattern shown here.

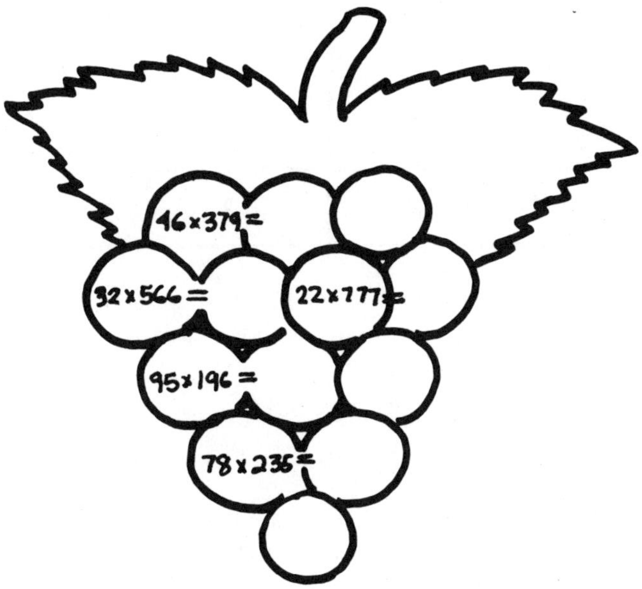

2. Copy each of the following sets of problems onto a different bunch of grapes as shown in the illustration.

Set 1	*Set 2*	*Set 3*
46 × 379 =	52 × 485 =	55 × 328 =
32 × 566 =	65 × 376 =	62 × 298 =
22 × 777 =	42 × 557 =	89 × 198 =
95 × 196 =	39 × 638 =	26 × 666 =
78 × 235 =	27 × 858 =	57 × 422 =

3. Copy each of the following answers onto a different circle.

17,434	25,220	18,040
18,112	24,440	18,476
17,094	23,394	17,622
18,620	24,882	17,316
18,330	23,166	24,054

4. On the back of each bunch of grapes, write the problems from the front as well as the correct answers.

Game Play:

1. Place one of the bunch of grapes in front of you. Look through the circles for the correct answers to the problems. Place these "missing grapes" in their appropriate places.

2. When you are finished, turn over the bunch of grapes for the self-check.

Unlock It

Skill Reinforced: Single-digit divisor and three digit dividend/Regrouping/No remainders

Materials Needed:

- 10 sheets of 3″ × 7″ gray posterboard
- 10 sheets of 4″ × 5″ gray posterboard
- scissors
- felt-tipped pen
- color markers or stars

Construction Directions:

1. Cut and mark the 3″ × 7″ posterboard pieces using the key pattern shown here.

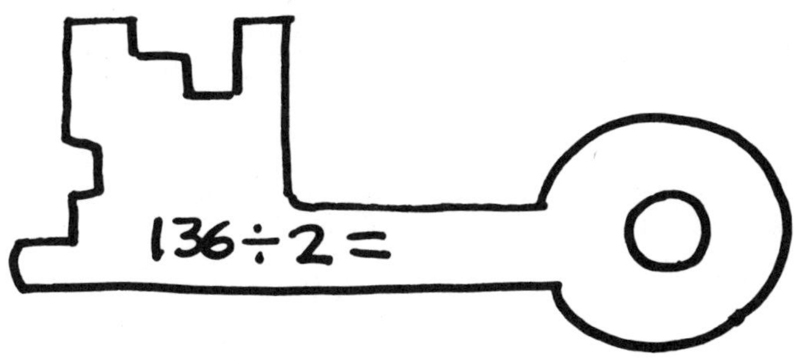

2. Cut and mark the 4″ × 5″ posterboard pieces using the lock pattern shown below.

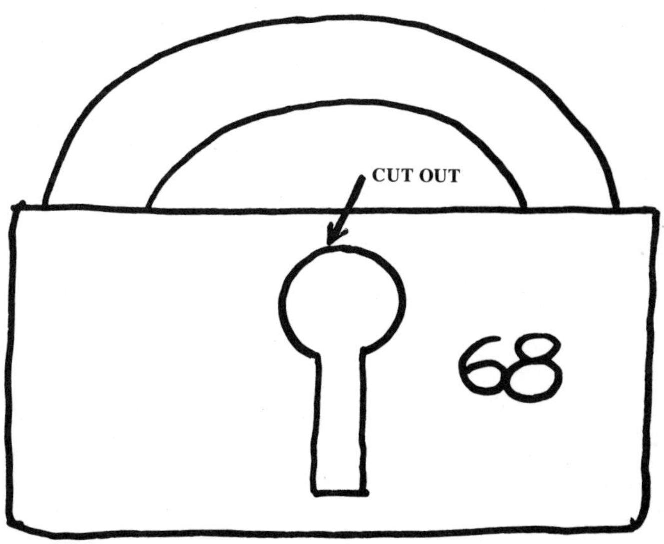

3. Copy the following problems onto the keys and the answers onto the locks as shown in the illustrations. Mark the backs of the corresponding pieces with the same-color markers or stars for the self-check.

136 ÷ 2 = 68	176 ÷ 4 = 44
174 ÷ 6 = 29	147 ÷ 3 = 49
165 ÷ 5 = 33	168 ÷ 7 = 24
144 ÷ 9 = 16	256 ÷ 8 = 32
102 ÷ 6 = 17	196 ÷ 7 = 28

Game Play:
1. Place all the keys in one pile and all the locks in another. Be sure that the keys have the problem side facing up.
2. Look at the problem on one of the keys. Do you know the answer? Look through the locks until you find it. Place the key in that lock. Do the same with all the rest of the game pieces.
3. When you are finished, turn over the pieces for the self-check. If you are correct, each lock and key pair will have the same-color marker on the back.

You Can Be a Rembrandt!

Skill Reinforced: Double-digit divisor and multidigit dividend/Regrouping/Remainders

Materials Needed:
- 5 sheets of 10″ × 12″ white posterboard
- 5 large pictures from magazines
- scissors
- felt-tipped pen
- glue

Construction Directions:
1. Glue a different picture to each of the posterboard pieces as shown here. Leave an inch of the posterboard exposed on the bottom, where you will write out the problem and answer.

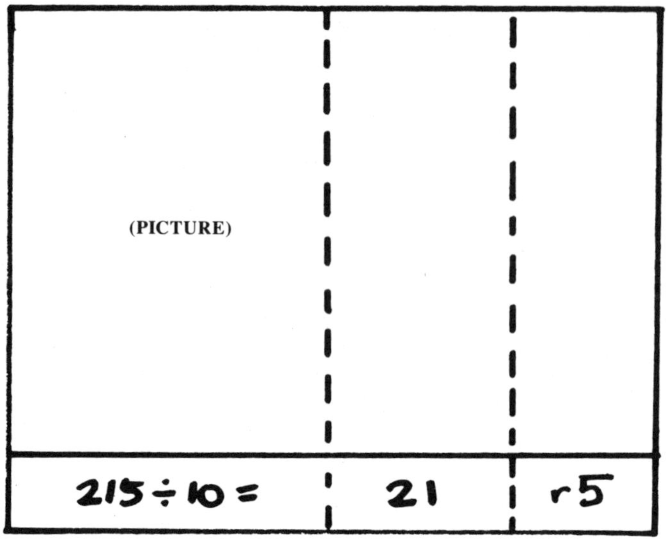

2. Copy each of the following problems and answers onto a different posterboard as shown in the illustration.

215 ÷ 10 = 21 r5

3423 ÷ 30 = 114 r3

1162 ÷ 20 = 58 r2

4081 ÷ 40 = 102 r1

2104 ÷ 50 = 42 r4

3. Cut the problem and picture apart where indicated by the dotted lines.

Game Play:
1. Place the puzzle pieces with problems in one pile, those with answers in another pile, and those with remainders in still another.
2. Take one of the puzzle pieces with problems and look through the other two piles for the correct answer and remainder. When you find them, place them next to the problem. If you are correct, you will see a complete picture.
3. Do this with all the puzzle pieces. You can form five pictures.

PART II
Games for Group Instruction

Participation in mathematics games and activities, in both large and small groups, promotes the acquisition of skills and concepts and provides motivation to learn. Additionally, games offer the kind of variety that makes drill and practice an enjoyable experience instead of one that is rather routine and dull.

The use of group games in mathematics instruction provides a beneficial exchange of ideas and a sharing of learning among students of all ability levels. The game method of reinforcement keeps students of high ability from becoming bored with "the same old worksheets," it reinforces students of average ability, and it encourages students of lesser ability to keep trying to achieve. Moreover, teachers can utilize mathematics game-playing as a diagnostic tool to determine the need for either challenge or remediation among students.

The group games, like the individual games, are easy to make, and each includes clear construction directions and a list of needed materials. The games are arranged in mathematical sequence from the easiest skills to the more difficult.

In the Everglades

Skill Reinforced: Multidigit addends/Multiple regrouping

Materials Needed:
- 3 egg cartons
- 18 sheets of 3″ × 4″ blue posterboard
- 18 sheets of 2″ × 2½″ green posterboard
- scissors
- felt-tipped pen

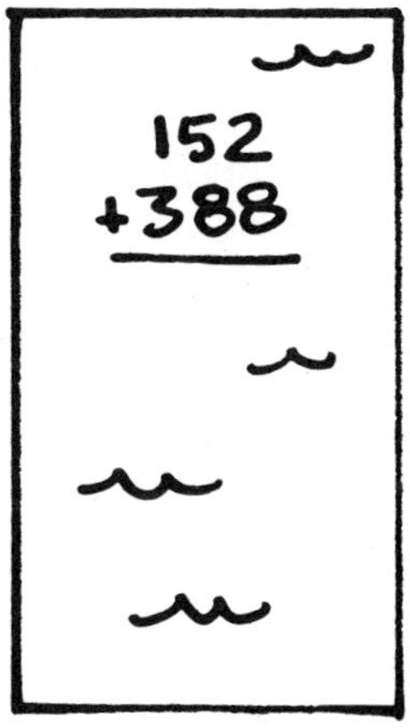

Construction Directions:
1. Cut the egg cartons lengthwise down the center so that there is a row of six humps when the half is turned over. Now cut a crosswise slit in each hump.
2. Mark the blue cards as shown here and copy each of the following problems onto a different card.

152 +388	267 +173	426 +187	159 +256	618 +193
167 +165	392 +168	294 +326	119 +292	393 +107
456 +169	365 +477	623 +198	166 +248	278 +173
565 +178	777 +129	528 +184		

3. Mark the green cards as shown here and copy each of the following answers onto a different card.

540, 440, 613, 415, 811, 332, 560, 620, 411, 500, 625, 842, 821, 414, 451, 743, 906, 712

Game Play: This game may be played by two or three players.

1. Each player selects an egg-carton gameboard and places it upside down in front of her.

2. One player deals out six blue cards to each player to place in the egg-carton holder, one card per slit. The green cards are placed face down in the center of the playing area.

3. The first player takes the top green card and sees if the answer on it fits any of the six problems on her egg-carton holder. If it does, she puts the green card in the same slit that holds the corresponding blue problem card.

4. If the player is unable to match the answer to any of her problems, she places the green card (problem side up) next to the stack of green cards.

5. The next player continues the game in the same manner.

6. The first player to match six answers to their correct problems is the winner. If all the green cards in the original stack have been picked and there is still no winner, the green cards that have not been matched are turned over and the game continues.

Trawling for a Good Catch

Skill Reinforced: Multidigit subtrahend and multidigit minuend/Regrouping

Materials Needed:

- 4 sheets of 11″ × $8\frac{1}{2}$″ white posterboard
- 32 sheets of $2\frac{1}{2}$″ × 3″ posterboard—8 sheets each of yellow, green, red, and orange
- scissors
- felt-tipped pens in various colors

Construction Directions:

1. Mark the large posterboard pieces using an enlargement of the gameboard pattern shown here.

2. Cut and mark the small posterboard pieces using the fish pattern shown here.

3. Copy each of the following answers onto a different color fish in the manner shown in the illustration.

539, 548, 749, 306, 204, 227, 253, 108

Game Play: This game is for two to four players.
1. Each player selects a gameboard and eight fish of the same color.
2. When all the players are ready, one player says "begin." At this signal, each player attempts to place all his fish on the gameboard, matching the problems with the correct answers. The first to do this is the winner.

Slide into First

Skill Reinforced: Double-digit multiplicand and single-digit multiplier/Multiples of ten

Materials Needed:
- 4 sheets of $2\frac{1}{2}'' \times 10''$ brown posterboard
- 16 circles of white posterboard with $1\frac{1}{2}''$ diameters
- scissors
- felt-tipped pen

Construction Directions:
1. Cut and mark the brown posterboard pieces using an enlargement of the bat pattern shown here.

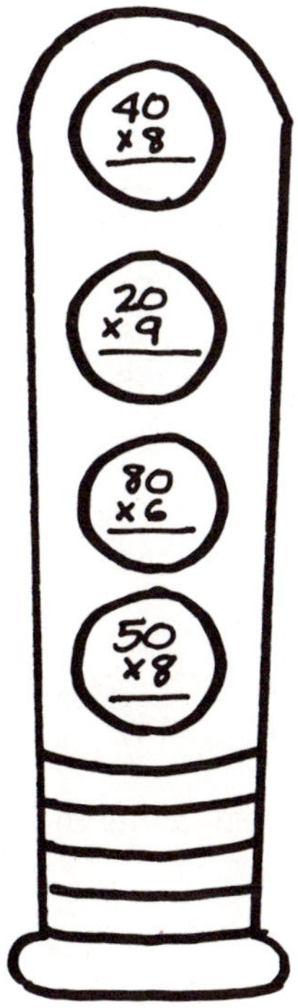

2. Copy each of the following groups of problems onto a different bat in the manner shown in the illustration.

Group 1

40	20	80	50
× 8	× 9	× 6	× 8

Group 2

60	30	90	70
× 7	× 8	× 3	× 7

Group 3

50	80	40	60
× 5	× 7	× 5	× 9

Group 4

70	90	20	80
× 9	× 5	× 7	× 8

3. Mark the posterboard circles using the baseball pattern shown here.

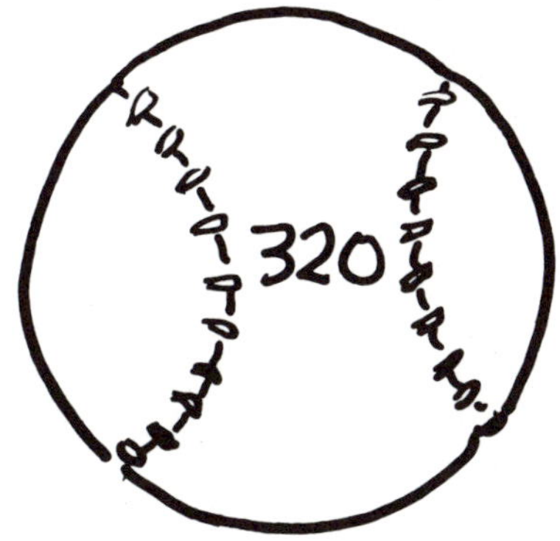

4. Copy each of the following answers onto a different baseball as shown in the illustration.

320, 180, 480, 400, 420, 240, 270, 490,
250, 560, 200, 540, 630, 450, 140, 640

Game Play: This game is for four players.
1. Each player chooses a bat gameboard and places it in front of her.
2. All the baseballs are dealt face down to the players. Players should not be able to see the answers on each other's baseballs.
3. Players now look at the baseballs in their hands. If they have any answers to the problems on their bat, they place these balls in the appropriate space.
4. The player who dealt the baseballs now begins the game by holding out his baseballs, face down, to the player on his left. This player takes a baseball, being careful not to show it to any other player. If she can make a match, she places it in the appropriate space. If not, she simply adds it to the baseballs in her hand and continues the game by holding out her baseballs, face down, to the player on her left.
5. The game continues in this manner until one of the players covers all four problems on his or her gameboard. This player is the winner.

Blast Off

Skill Reinforced: Double-digit multiplicand and single-digit multiplier/No regrouping

Materials Needed:
- 5 sheets of 8″ × 8″ corrugated cardboard
- 25 sheets of 3″ × 6″ red posterboard
- scissors
- felt-tipped pen

Construction Directions:

1. Cut five $2\frac{1}{2}$″ slits in the corrugated gameboards as shown here.

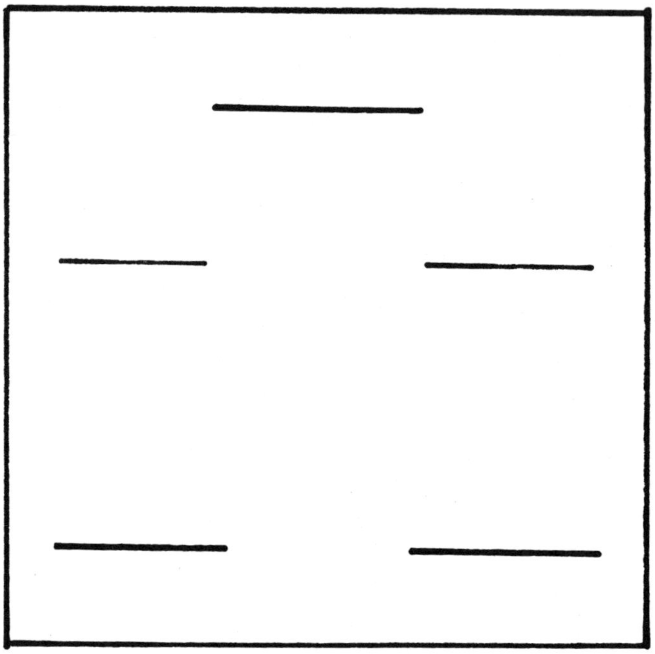

2. Mark and cut the red posterboard pieces using an enlargement of the rocket pattern shown above.

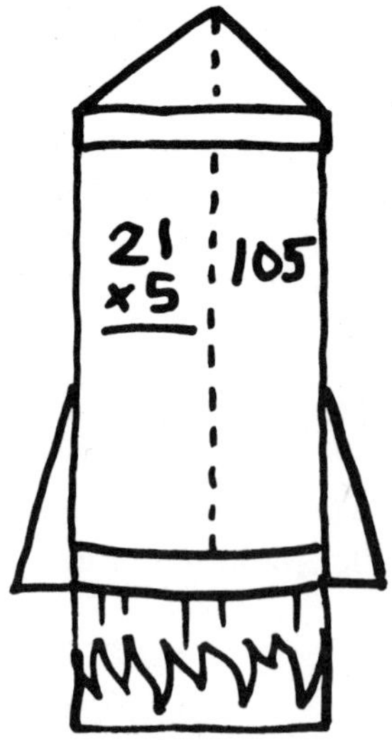

3. Copy each of the following problems and answers onto a different rocket in the manner shown in the illustration. (The problem should be on one side and the answer on the other.)

21	31	32	42	41
× 5	× 6	× 4	× 3	× 7
105	186	128	126	287
51	52	53	54	61
× 8	× 4	× 3	× 2	× 7
408	208	159	108	427
62	63	71	72	81
× 4	× 3	× 5	× 4	× 6
248	189	355	288	486
82	83	84	91	92
× 4	× 3	× 2	× 3	× 4
328	249	168	273	368
93	31	43	73	41
× 2	× 9	× 3	× 3	× 4
186	279	129	219	164

4. Cut the rockets in half along the dotted line.

Game Play: This game is for two to five players.
1. Separate the rocket pieces and place the right sides of the rockets in one stack and the left sides in another. All playing pieces should be placed face down in the center of the playing area. Each player receives a heavy gameboard and then selects five left sides of the rocket without looking at the problems. These are placed in each of the five slots in the gameboards. Players must not show their rockets to each other.
2. All the right halves are now dealt out face down to the players.
3. Players should now look through the rocket halves in their hands and try to match them with the halves on their gameboard. If they are able to match the problems with the correct answers, they should fit these together in their gameboards.
4. The dealer begins by holding out his unmatched rocket halves, face down, to the player on his left. This player must take one of the pieces, being careful not to show it to any other player. If it is a match, she places it in the appropriate slot. If not, she simply adds it to the playing pieces already in her hand and continues the game by holding out her rocket halves, face down, to the player on her left.
5. The game continues in this manner until one of the players has all five correctly matched rockets. This player is the winner.

As Easy as ABC!

Skill Reinforced: Single-digit divisor and double-digit dividend/No regrouping/No remainders

Materials Needed:
- 1 ditto master
- 1 clear transparency
- 9 markers for each player (paper clips or corn kernels work fine)

Construction Directions:
1. Draw the following ABC gameboard on the ditto master and make a copy for each player.

A	B	C

2. Copy the following list of problems onto the transparency.

$26 \div 2 =$	$48 \div 2 =$	$82 \div 2 =$
$33 \div 3 =$	$96 \div 3 =$	$50 \div 5 =$
$88 \div 2 =$	$48 \div 4 =$	$80 \div 2 =$
$84 \div 4 =$	$60 \div 3 =$	$86 \div 2 =$
$99 \div 3 =$	$28 \div 2 =$	$93 \div 3 =$
$68 \div 2 =$	$44 \div 2 =$	$90 \div 3 =$
$69 \div 3 =$	$80 \div 2 =$	

Game Play: This game may be played by any number of players.
1. Pass out an ABC gameboard and nine markers to each player.
2. Show the problem list on the transparency to the players and explain that they are to select nine problems from the list and write a different one in each of the squares on their gameboard.

3. You now begin the game by reading through the following answers, one by one. Each player looks over his gameboard (in bingo fashion) trying to locate the problems whose answers match those you read. When a player finds one, she places a marker in that square. The game continues in this manner until someone gets three markers in a row. The row may be vertical, horizontal, or diagonal. This player says "Easy as ABC" and is the winner. Be sure to check to see if the players' marked squares have actually been called.

(read these)	*(correct problems)*
13	26 ÷ 2 =
24	48 ÷ 2 =
41	82 ÷ 2 =
11	33 ÷ 3 =
32	96 ÷ 3 =
10	50 ÷ 5 =
44	88 ÷ 2 =
12	48 ÷ 4 =
40	80 ÷ 2 =
21	84 ÷ 4 =
20	60 ÷ 3 =
43	86 ÷ 2 =
33	99 ÷ 3 =
14	28 ÷ 2 =
31	93 ÷ 3 =
14	28 ÷ 2 =
31	93 ÷ 3 =
34	68 ÷ 2 =
22	44 ÷ 2 =
30	90 ÷ 3 =
23	69 ÷ 3 =
40	80 ÷ 2 =

Watch Out for Cecil!

Skill Reinforced: Double-digit dividend and single-digit divisor/Remainders

Materials Needed:
- 21 sheets of $2\frac{1}{2}'' \times 3\frac{1}{2}''$ yellow posterboard
- 20 sheets of $2\frac{1}{2}'' \times 3\frac{1}{2}''$ red posterboard
- felt-tipped pen

Construction Directions:

1. Copy each of the following problems onto a different yellow card.

37 ÷ 6 =	89 ÷ 9 =
87 ÷ 9 =	63 ÷ 8 =
43 ÷ 5 =	12 ÷ 7 =
26 ÷ 3 =	17 ÷ 3 =
27 ÷ 8 =	40 ÷ 9 =
17 ÷ 6 =	80 ÷ 9 =
41 ÷ 6 =	35 ÷ 8 =
66 ÷ 7 =	47 ÷ 6 =
44 ÷ 9 =	39 ÷ 7 =
71 ÷ 8 =	18 ÷ 7 =

2. Copy each of the following answers onto a different red card.

6 r1	9 r8
9 r6	7 r7
8 r3	1 r5
8 r2	5 r2
3 r3	4 r4
2 r5	8 r8
6 r5	4 r3
9 r3	7 r5
4 r8	5 r4
8 r7	2 r4

3. Copy the following onto one of the yellow cards.

Game Play: This game is for two to four players.
1. The dealer hands out all the cards. Players should not let each other see their cards.
2. Players look through their cards searching for answers on their red cards that match the problems on their yellow cards.
3. The dealer begins by holding out his cards face down to the player on his left, who takes a card. She should not show it to any other player. If it matches another of her cards, she lays them both down. If not, she adds it to the cards in her hand and continues the game by holding out the cards to the player on her left.
4. The game continues until all cards are paired. The player left holding the "Cecil" card is the loser and receives a "C." When a player has accumulated all the letters in the word "Cecil," he is out of the game. The last player remaining is the winner.

Frantic Football

Skill Reinforced: Multidigit multiplicand and single-digit multiplier/Regrouping

Materials Needed:
- 1 sheet of 8″ × 16″ green posterboard
- 1 sheet of $2\frac{1}{2}$″ × 3″ red posterboard
- 1 sheet of $2\frac{1}{2}$″ × 3″ yellow posterboard
- 30 sheets of 2″ × 3″ white posterboard
- scissors
- felt-tipped pen

Construction Directions:
1. Mark the green posterboard using an enlargement of the football-field pattern shown here.

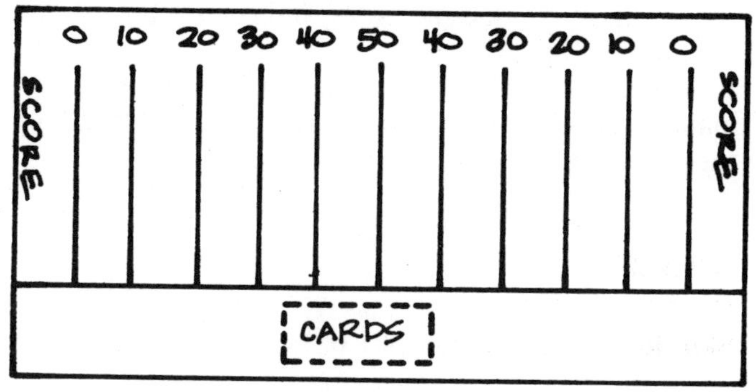

2. Cut and mark both the red and yellow posterboard pieces using the football pattern below.

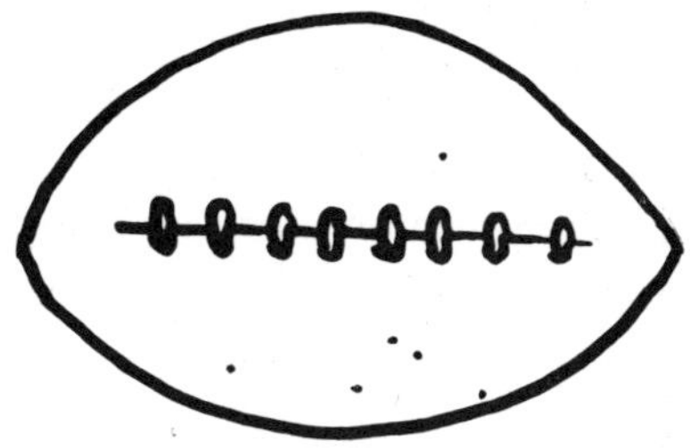

3. Copy each of the following problems onto a different white card.

112 × 6	113 × 5	114 × 4	115 × 3	116 × 2
121 × 6	122 × 5	123 × 4	125 × 3	127 × 2
132 × 6	134 × 5	136 × 4	138 × 3	142 × 5
145 × 4	147 × 3	149 × 2	152 × 5	154 × 4
158 × 3	163 × 4	165 × 3	169 × 2	174 × 4
176 × 3	178 × 2	183 × 4	185 × 3	189 × 2

Game Play: This game is for two players.
1. Each player chooses a football and places it on one of the ten-yard lines.
2. The white problem cards are placed face down in the area indicated on the gameboard.

3. The player with the red football goes first. She takes the top card and reads the problem aloud. If she can give the correct answer, she moves her football ten yards. If not, she does not move and returns the card to the bottom of the pile.
4. The next player continues the game in the same manner. The first player who moves across the length of the gameboard and scores is the winner.

Be a Winner

Skill Reinforced: Multiples of 1000

Materials Needed:
- 3 sheets of 6″ × 18″ white posterboard
- 18 sheets of 3″ × 3″ red posterboard
- scissors
- felt-tipped pen

Construction Directions:
1. Mark the large posterboard pieces using an enlargement of the gameboard pattern shown here.

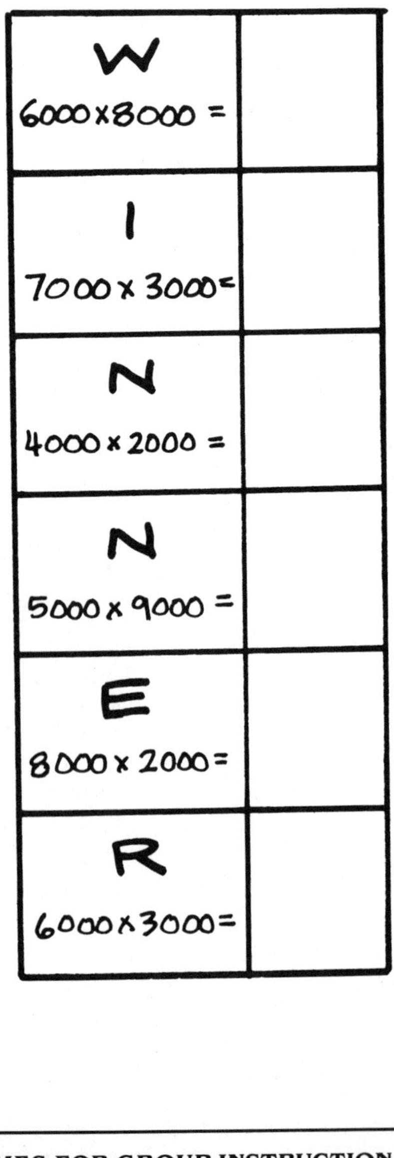

2. Copy each of the following problem groups onto a different gameboard in the manner shown in the illustration.

Problem Group 1	***Problem Group 2***
6000 × 8000 =	2000 × 5000 =
7000 × 3000 =	4000 × 8000 =
4000 × 2000 =	3000 × 9000 =
5000 × 9000 =	4000 × 5000 =
8000 × 2000 =	6000 × 7000 =
6000 × 3000 =	2000 × 7000 =

Problem Group 3
3000 × 8000 =
9000 × 7000 =
8000 × 8000 =
2000 × 6000 =
8000 × 9000 =
3000 × 2000 =

3. Cut the small posterboard pieces using the following letter patterns. Cut as many as indicated on each letter.

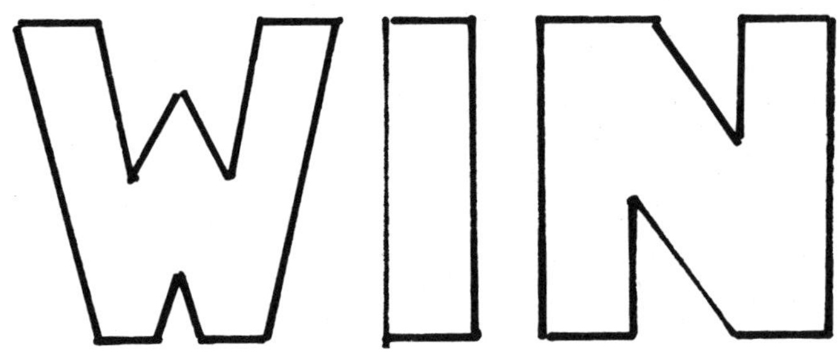

(3 LETTERS) (3 LETTERS) (6 LETTERS)

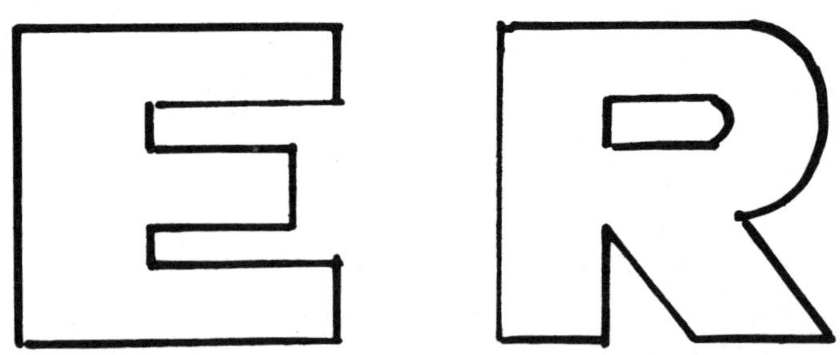

(3 LETTERS) (3 LETTERS)

4. Copy each of the following answers onto a different letter, as indicated.

Letter W	*Letter I*
48,000,000	21,000,000
10,000,000	32,000,000
24,000,000	63,000,000

Letter E	*Letter R*
16,000,000	18,000,000
42,000,000	14,000,000
72,000,000	6,000,000

Letter N
8,000,000
45,000,000
27,000,000
20,000,000
64,000,000
12,000,000

Game Play: This game is for two or three players.
1. Each player takes a gameboard. The letters are separated and placed into stacks, same letters together, in between the players.
2. The first player picks the top letter from one of the stacks. If the answer matches the problem next to that letter on his gameboard, he places the letter in the appropriate space. If not, he places it face down on the bottom of the original stack. In either case, the player to his left takes the next turn.
3. The game continues in this manner until a player correctly fills all the spaces on his gameboard. This player is the winner.

Pirate's Pistol

Skill Reinforced: Double-digit multiplicand and double digit multiplier/Regrouping

Materials Needed:
- 20 sheets of 3″ × 4″ yellow posterboard
- 20 sheets of 3″ × 4″ red posterboard
- felt-tipped pen

Construction Directions:
1. Mark each of the yellow cards using the pattern shown here.

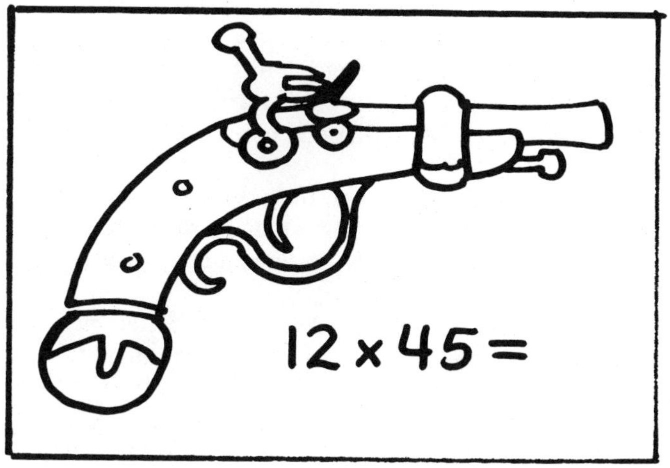

2. Mark each of the red cards using the pattern shown here.

3. Copy each of the following problems onto a different yellow card as shown in the illustration.

12 × 45 =	23 × 38 =	28 × 32 =
16 × 42 =	15 × 43 =	26 × 34 =
25 × 37 =	14 × 44 =	16 × 46 =
27 × 36 =	22 × 38 =	14 × 45 =
13 × 47 =	23 × 37 =	29 × 32 =
18 × 48 =	19 × 43 =	26 × 33 =
23 × 35 =	12 × 49 =	

4. Copy each of the following answers onto a different red card as shown in the illustration.

540, 874, 896, 672, 645, 884, 925, 616, 736, 972, 836, 630, 611, 851, 928, 864, 817, 858, 805, 588

Game Play: This game is for four to six players.

1. A player shuffles all the cards together and deals them out. Players should not let other players see their cards.

2. The players look through their cards to find problems on their yellow cards that correspond to answers on their red cards. If they find any, they lay both the red and yellow cards down.

3. The dealer begins by holding out her cards, face down, to the player on her left. This player must take one of the cards, being careful not to show it to anyone. If any of her red cards can be matched to a yellow card, she lays them both down. If not, she keeps the card in her hand and continues the game by holding out her cards to the player on her left.

4. The game continues until all cards are matched. The first player to match all of his cards is the winner.

Celebrate the Fourth

Skill Reinforced: Double-digit divisor and multidigit dividend/Multiples of ten/No regrouping/No remainders

Materials Needed:

- 5 sheets of 12″ × 12″ white posterboard
- 20 sheets of 2″ × 5″ red posterboard
- scissors
- felt-tipped pens in various colors
- heavy twine
- tape

Construction Directions:

1. Mark and cut four $2\frac{1}{2}$″ slits in the five white posterboard pieces using an enlargement of the gameboard pattern shown here.

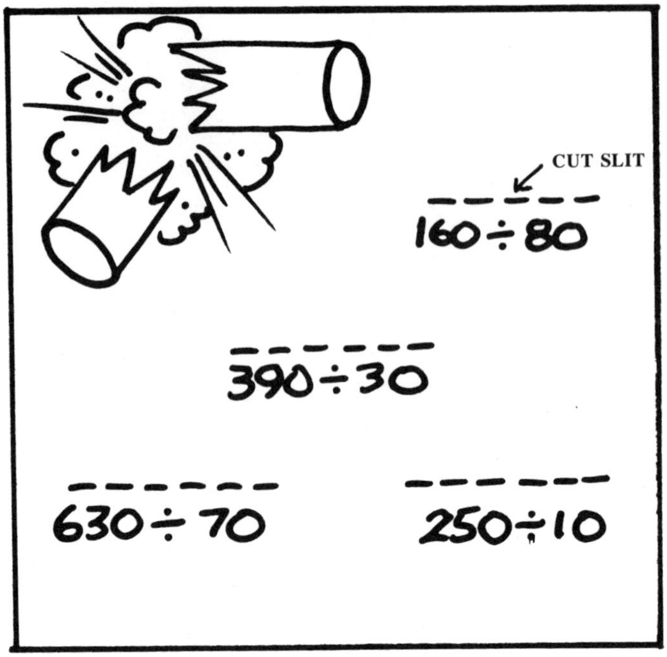

2. On each of the gameboards copy one of the following problem groups. Each problem should be copied beneath a slit, as shown in the illustration.

Group 1	*Group 2*	*Group 3*
160 ÷ 80	240 ÷ 80	360 ÷ 90
390 ÷ 30	900 ÷ 90	550 ÷ 50
630 ÷ 70	240 ÷ 30	420 ÷ 20
250 ÷ 10	480 ÷ 20	150 ÷ 10

Group 4	*Group 5*
250 ÷ 50	540 ÷ 90
560 ÷ 80	120 ÷ 10
280 ÷ 20	400 ÷ 20
660 ÷ 30	690 ÷ 30

3. Cut and mark the red posterboard pieces using the firecracker pattern shown here. Attach a piece of twine using the tape as shown in the illustration.

4. Copy each of the following answers onto a different firecracker.

2, 3, 4, 5, 6, 7, 8, 9, 10, 11,
12, 13, 14, 15, 20, 21, 22, 23, 24, 25

Game Play: This game is for two to five players.
1. Each player chooses a gameboard. The firecrackers are placed face down in the center of the playing area.
2. The first player takes the top firecracker and looks over his gameboard to find a problem that produces the answer on the firecracker. If he finds one, he places the firecracker in the slit above the problem. If not, he places the firecracker face up next to the original stack.
3. The next player continues the game in the same manner. The first one to fill her gameboard is the winner. If the players use all the firecrackers in the original stack and there is still no winner, the firecrackers not on the gameboards are turned over and the game continues.

Hit a Homer

Skill Reinforced: Double-digit divisor and multidigit dividend/Regrouping/Remainders

Materials Needed:
- 4 sheets of 9″ × 12″ tan posterboard
- 10 white posterboard circles with 3″ diameters
- 10 yellow posterboard circles with 3″ diameters
- 1 sheet of white construction paper
- 1 sheet of yellow construction paper
- glue
- scissors
- felt-tipped pen
- spinner from an old game

Construction Directions:
1. Mark the tan posterboard pieces using an enlargement of the baseball-field pattern shown here.

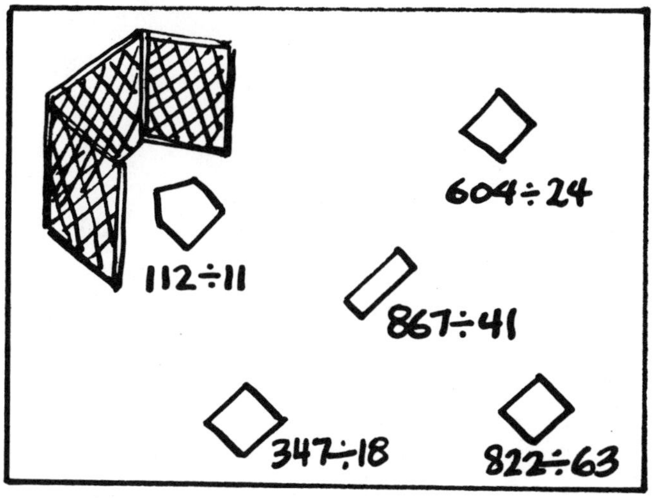

2. Copy each group of problems onto a different tan gameboard in the manner shown in the illustration.

Group 1	*Group 2*
112 ÷ 11	159 ÷ 13
347 ÷ 18	422 ÷ 21
604 ÷ 24	708 ÷ 26
867 ÷ 41	389 ÷ 24
822 ÷ 63	924 ÷ 92

Group 3	*Group 4*
211 ÷ 14	514 ÷ 16
997 ÷ 31	360 ÷ 21
816 ÷ 28	856 ÷ 37
594 ÷ 42	626 ÷ 52
184 ÷ 12	908 ÷ 82

3. Mark all the circles using the baseball pattern shown here.

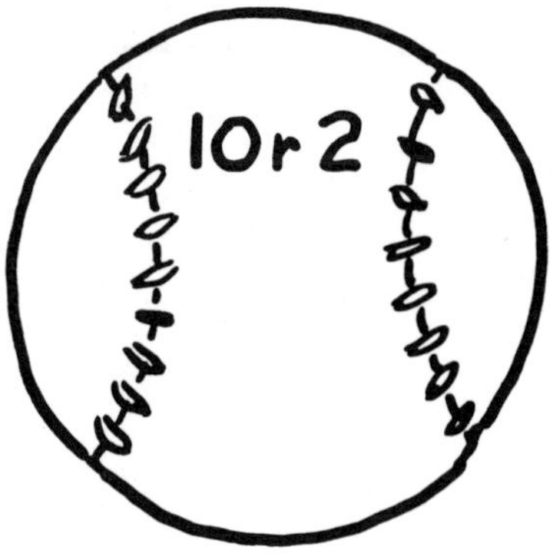

4. Copy each of the following answers onto a different white baseball as shown in the illustration.

10 r2	12 r3
19 r5	20 r2
25 r4	27 r6
21 r6	16 r5
13 r3	10 r4

5. Copy each of the following answers onto a different yellow baseball.

15 r1	32 r2
32 r5	17 r3
29 r4	23 r5
14 r6	12 r2
15 r4	11 r6

6. Cut and glue the sheets of white and yellow construction paper to the spinner as shown here.

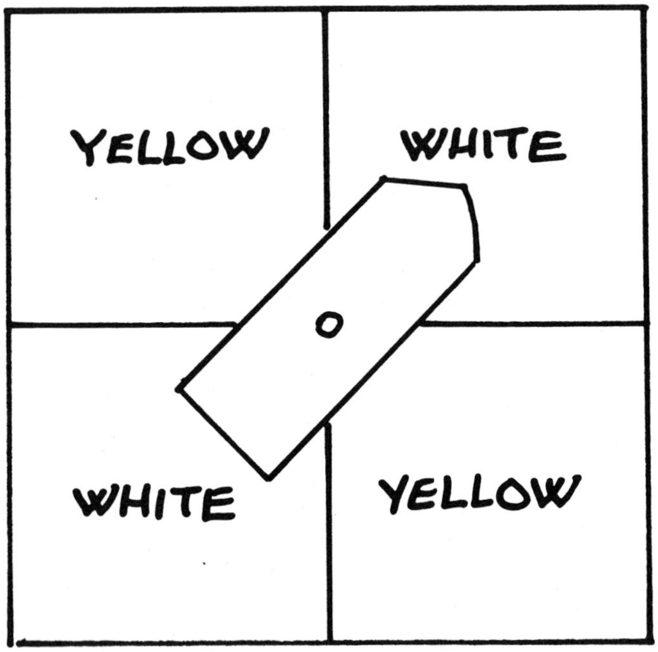

Game Play: This game is for two to four players.

1. Each player chooses a tan gameboard. The yellow and white baseballs are separated according to color and placed numeral side down in the corner of the playing area.

2. The first player spins. She then picks up the top baseball from either the yellow or white pile, depending on the spin.

3. Now she looks over the problems on her gameboard to find one that produces the answer on her baseball. If the player is able to do this, she places the baseball on top of the baseball diamond on the gameboard. If not, she places the baseball, answer side up, next to the original pile of baseballs.

4. The next player continues the game in the same manner. The first one to place five baseballs on his gameboard is the winner.

5. If the players use all the baseballs in either the yellow or white stack and there is still no winner, the baseballs not on the gameboards are turned over and the game continues.

PART III

Puzzle Sheets

The puzzle sheets described in this part of the book lead students to the solutions of problems and equations in a unique and challenging way. The puzzles are, for the most part, self-checking, to give immediate feedback to both students and teachers. For those puzzles that are not self-checking, answer keys are included for quick and easy reference.

The puzzle sheets are grouped together at the back of the book so that they may be detached and duplicated with a minimum of time and effort. In this way, they may be put to use most conveniently as reinforcers of a variety of mathematical skills.

The Big Challenge

Skill Reinforced: Identifying properties of addition and multiplication

Puzzle Description: The students are to match a list of number sentences with a list of properties. As they do so, they are to write the circled letter in each property next to its corresponding number sentence. If they are correct, the students will spell how well they did. This puzzle is on page 46.

Answer Key: The puzzle is self-correcting. As shown below, the circled letters will spell the words SUPER STAR if the puzzle is completed correctly.

S	a(s)sociative-addition
U	comm(u)tative-multiplication
P	identify-multi(p)lication
E	id(e)ntity-addition
R	distributive-multiplication ove(r) addition
S	as(s)ociative-multiplication
T	inverse-addi(t)ion
A	commutative-(a)ddition
R	inve(r)se-multiplication

Down Killer Hill!

Skill Reinforced: Multidigit subtrahend, multidigit minuend/Regrouping tens as ones

Puzzle Description: This puzzle consists of a drawing of a sled beginning to descend a steep hill. Along the way are problems. The answers to these problems may be found in the snowman. The students are to mark the appropriate answer/letter pairs next to each problem. If students answer correctly, the letters will spell out how long it took the sled to coast down the hill. This puzzle is on page 47.

Answer Key: The puzzle is self-correcting. As shown below, A MINUTE will be spelled if the puzzle is completed correctly.

A	109
M	145
I	136
N	164
U	127
T	158
E	117

Trip Through Space

Skill Reinforced: Double-digit multiplicand, single-digit multiplier/Regrouping

Puzzle Description The puzzle consists of a drawing of a space ship and its path through outer space. Along the way are ten stars; inside each of these is a problem. The answers may be found in the planet. The students are to copy the correct answer/letter pairs next to each problem. If they are correct, the letters will spell where the space ship is traveling. The puzzle may be found on page 48.

Answer Key: The puzzle is self-correcting. As shown below, OUTER SPACE will be spelled if the puzzle is completed correctly.

O	94
U	87
T	92
E	95
R	90
S	98
P	96
A	85
C	84
E	81

You'll Never Guess

Skill Reinforced: Double-digit dividend, single-digit divisor

Puzzle Description: The puzzle consists of one list of problems and another of answers. Students match the correct answers to the corresponding problem by writing the answer/letter pairs in the appropriate spaces. If students complete the puzzle correctly, the letters will spell what's in the birthday package. This puzzle is on page 49.

Answer Key: The puzzle is self-correcting. As shown below, ROLLERSKATES will be spelled if the puzzle is completed correctly.

R	5
O	3
L	4
L	11
E	2
R	8
S	7
K	6
A	9
T	13
E	14
S	10

Wanda's Rocket

Skill Reinforced: Base 5 addition

Puzzle Description: The students are to cut out the answer cards and place them over the correct problems as they follow Wanda's rocket. If they are correct, the letters on the answer cards will spell how Wanda feels as her rocket appears to be headed for a crash-landing. The puzzle is on page 50.

Answer Key: The puzzle is self-correcting. As shown below, the letters will spell SCARED if the puzzle is completed correctly.

S	123
C	1110
A	1224
R	443
E	122
D	1003

Runaway Robot

Skill Reinforced: Subtraction of integer

Puzzle Description: The students follow the robot's path. Each time they come to a problem that has a correct answer, they copy down the letter under it on one of the spaces at the bottom of the page. If they are right, these letters will spell the robot's name. This puzzle is found on page 51.

Answer Key: The puzzle is self-correcting. As shown below, the letters under the correct equations will spell RONALD MCMETAL.

R	$4 - 6 = -2$
O	$-5 - 12 = -17$
N	$-3 - (-8) = 5$
A	$7 - (-2) = 9$
L	$8 - 8 = 0$
D	$-2 - 5 = -7$
M	$-8 - 2 = -10$
C	$-4 - 4 = -8$
M	$-10 - 12 = -22$
E	$-13 - 30 = -43$
T	$6 - (-1) = 7$
A	$5 - (-8) = 13$
L	$12 - (-4) = 16$

Dividing Animals

Skill Reinforced: Divisibility rules

Puzzle Description: This puzzle is actually three separate puzzles, each containing a list of numerals in a cage. The students are to look through the list in each cage and determine which numerals are divisible by the puzzle's number. They are to copy down the letters next to each divisible number in the spaces provided. If they are correct, the letters in each puzzle will spell a different kind of animal. The puzzles are on page 52.

Answer Key: The puzzle is self-correcting. As shown below, the words CAMEL, HORSE, and MONKEY will be spelled if the puzzle is completed correctly.

Puzzle 1:

1640	C
8925	A
2310	M
1585	E
1150	L

Puzzle 2:

3432	H
5924	O
1975	R
2988	S
6544	E

Puzzle 3:

3276	M
9060	O
1623	N
7128	K
4197	E
1101	Y

Cross Over the Bridge

Skill Reinforced: Multidigit dividend, double-digit divisor/Remainders

Puzzle Description: In this puzzle, students are to "cross over the bridge," beginning at the car. Along the way are problems. The answers to the problems are located in the ship. Students are to mark the appropriate answer/letter pairs next to each problem. If they are correct, the letters will spell the name of the bridge. The puzzle is on page 53.

Answer Key: The puzzle is self-correcting. As shown below, GOLDEN GATE will be spelled if the puzzle is completed correctly.

G	131 r1
O	216 r2
L	106 r5
D	121 r8
E	122 r6
N	105 r10
G	124 r7
A	110 r3
T	114 r9
E	209 r4

PART IV Learning Boards

This section of the book consists of display and bulletin board ideas that involve students in learning mathematics in an active way—a way that requires them to use more than just their visual sense. These boards not only encourage the development of specific mathematical skills or concepts but also create a high level of interest in the topic being studied.

Directions for constructing and using the boards are given in detail. All may be easily modified to fit the space available in any particular classroom.

The learning boards may be used to implement and/or enrich the textbook materials, whether they are used for introductory, developmental, or reinforcement purposes.

Bertha Broom Sweeps Clean

Skill Reinforced: Other number systems/Roman numerals

Materials Needed:

- yellow background paper
- tan posterboard
- felt-tipped pen
- thumbtacks
- small Jell-O boxes
- glue
- scissors
- stapler
- tan construction paper

Construction Directions:

1. Use an opaque projector to trace and cut the learning board's title broom from the tan posterboard. Attach the small Jell-O boxes to the board and glue the title broom to these. This will give a three-dimensional effect.

2. Cut and mark the tan posterboard using the broom pattern below. You will need eight brooms.

3. Copy each of the following numerals onto a different broom as shown in the illustration.

19, 42, 26, 68, 12, 77, 94, 35

4. Attach these to the board with the thumbtacks, as shown in the illustration.
5. If you wish, use an opaque projector to trace the Roman numerals.

6. Fold the tan construction paper to form an answer key and mark the front cover as shown. Write the following Roman numerals on the inside and attach the answer key to the board.

19	XIX	42	XLII
26	XXVI	68	LXVIII
12	XII	77	LXXVII
94	XCIV	35	XXXV

Learning Board Use: Students look at the numerals on the brooms and match them with the Roman numerals. They do this by swinging the brooms toward the appropriate Roman numeral. They may use the answer key when they are finished.

Newspaper Route

Skill Reinforced: Rounding three-digit numbers to the nearest tenth

Materials Needed:
- newspapers
- white construction paper
- red posterboard
- 12 paper-towel rolls
- tape
- scissors
- stapler
- thumbtacks
- felt-tipped pen
- glue
- 2 large grocery sacks

Construction Directions:

1. Attach the newspapers to the bulletin board for the background paper.

2. Use an opaque projector to trace the boy and the lettering onto the white construction paper, and attach this to the board.

3. Construct six mailboxes from the red posterboard in the manner shown here.

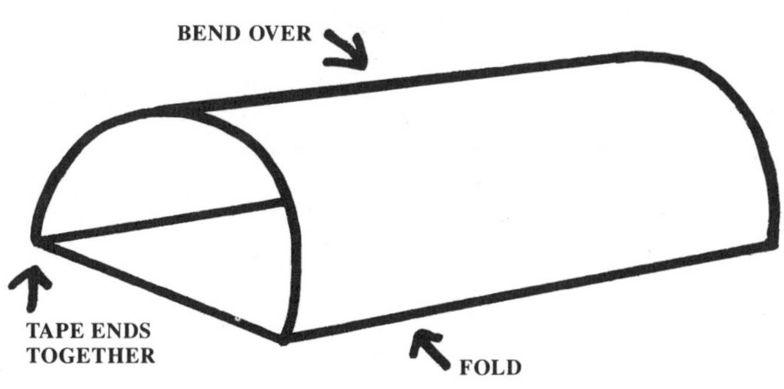

4. Copy each of the following numerals onto a different mailbox and attach to the board as shown in the illustration.

140, 150, 160, 170, 180, 190

5. Wrap each of the paper-towel rolls with newspaper and glue a piece of white construction paper to each as shown here.

6. Copy each of the following numerals onto a different newspaper roll as shown in the illustration.

136, 142, 147, 153, 158, 164,
166, 172, 177, 184, 187, 193

7. Place one paper sack inside the other, fold down their tops several folds, attach to the board, and place the newspaper rolls inside.
8. Fold the white construction paper and write "Answer Key" on the front cover. Write the following on the inside and attach the answer key to the board.

Mailbox	*Newspapers*
140	136, 142
150	147, 153
160	158, 164
170	166, 172
180	177, 184
190	187, 193

Learning Board Use: Students take each newspaper out of the sack and place it in the mailbox whose number represents the tenth nearest to the number on the newspaper. When they are finished, they may use the answer key to check themselves.

Score a Goal

Skill Reinforced: Ordering multidigit numbers

Materials Needed:
- white background paper
- felt-tipped pens in various colors
- straight pins
- gray posterboard
- small Jell-O boxes
- stapler
- glue
- blue construction paper
- legal-size envelope
- scissors

Construction Directions:
1. Use and opaque projector to trace and cut the hockey player from the white posterboard. Attach the Jell-O boxes to the board and glue the player to these. This will give a three-dimensional effect.
2. Use an opaque projector to trace the figures and lettering onto the background paper. Attach the straight pins as shown.
3. Cut and punch the gray posterboard using the hockey puck pattern shown below. You will need 12 pucks.
4. Copy each of the following numbers onto a different puck as shown in the illustration.

3192 5	3152 9	4678 23	6880 1	6888 8
4687 32	9269 9	9268 9	7985 8	5432 65
7958 5	5436 25			

Score A Goal

Directions

1. Take the hockey pucks out of the envelope and look at the numbers on them.
2. Starting at the hockey player, follow the hockey puck's path to the goal. Each time you come to a hook, place a puck on it. Just make sure that the pucks are in the correct numerical order.
3. Use the answer key when you finish.

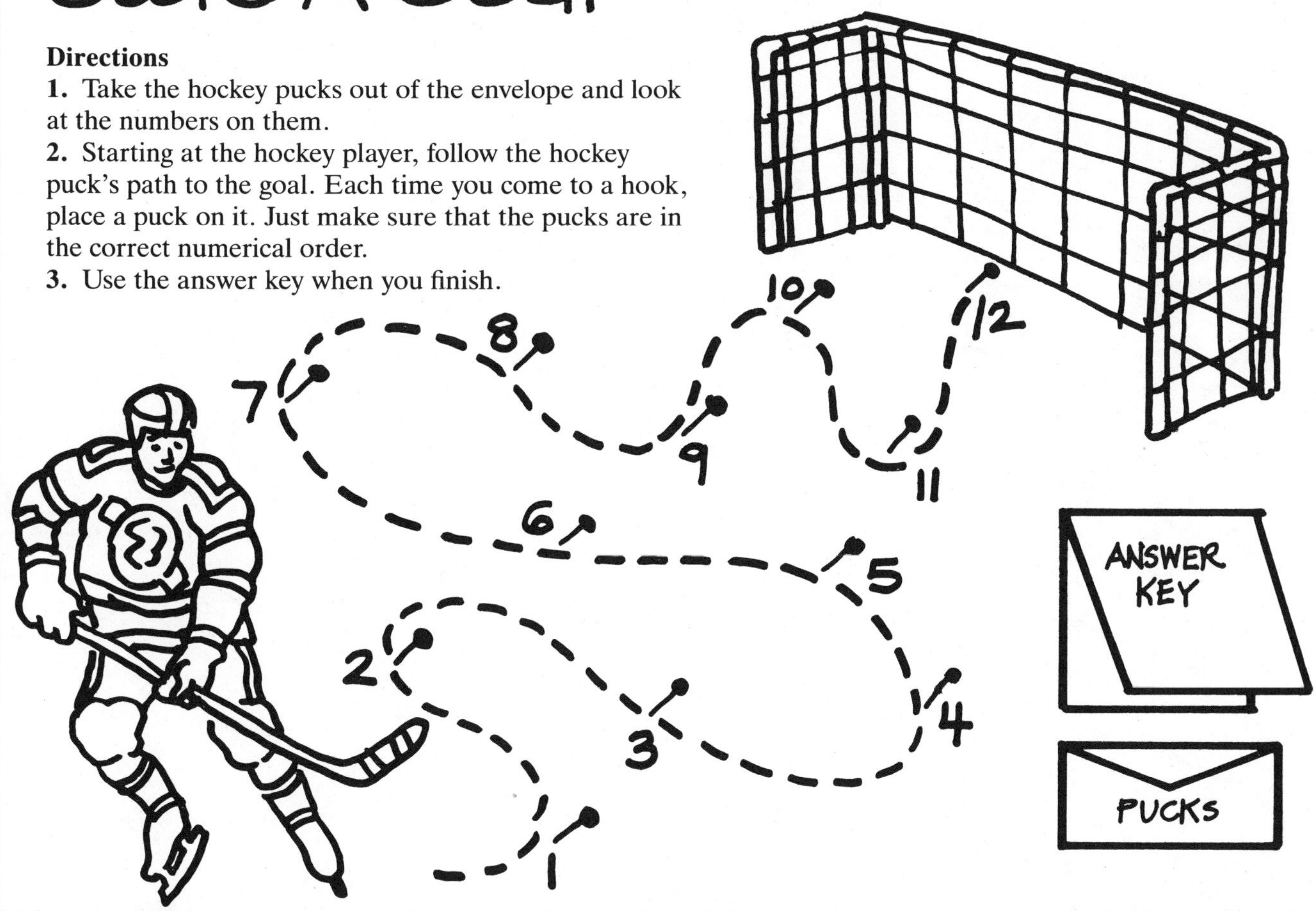

5. Write "Pucks" on the envelope, attach it to the board, and place the pucks in it.
6. Fold the blue construction paper to form an answer key and mark the front as shown in the board illustration. Copy the following onto the inside of the answer key.

31529
31925
467823
468732
543265
543625
68801
68888
79585
79858
92689
92699

Learning Board Use: Students take all the hockey pucks out of the envelope and hang them on the hooks in the correct numerical order. When they finish, they may use the answer key to check themselves.

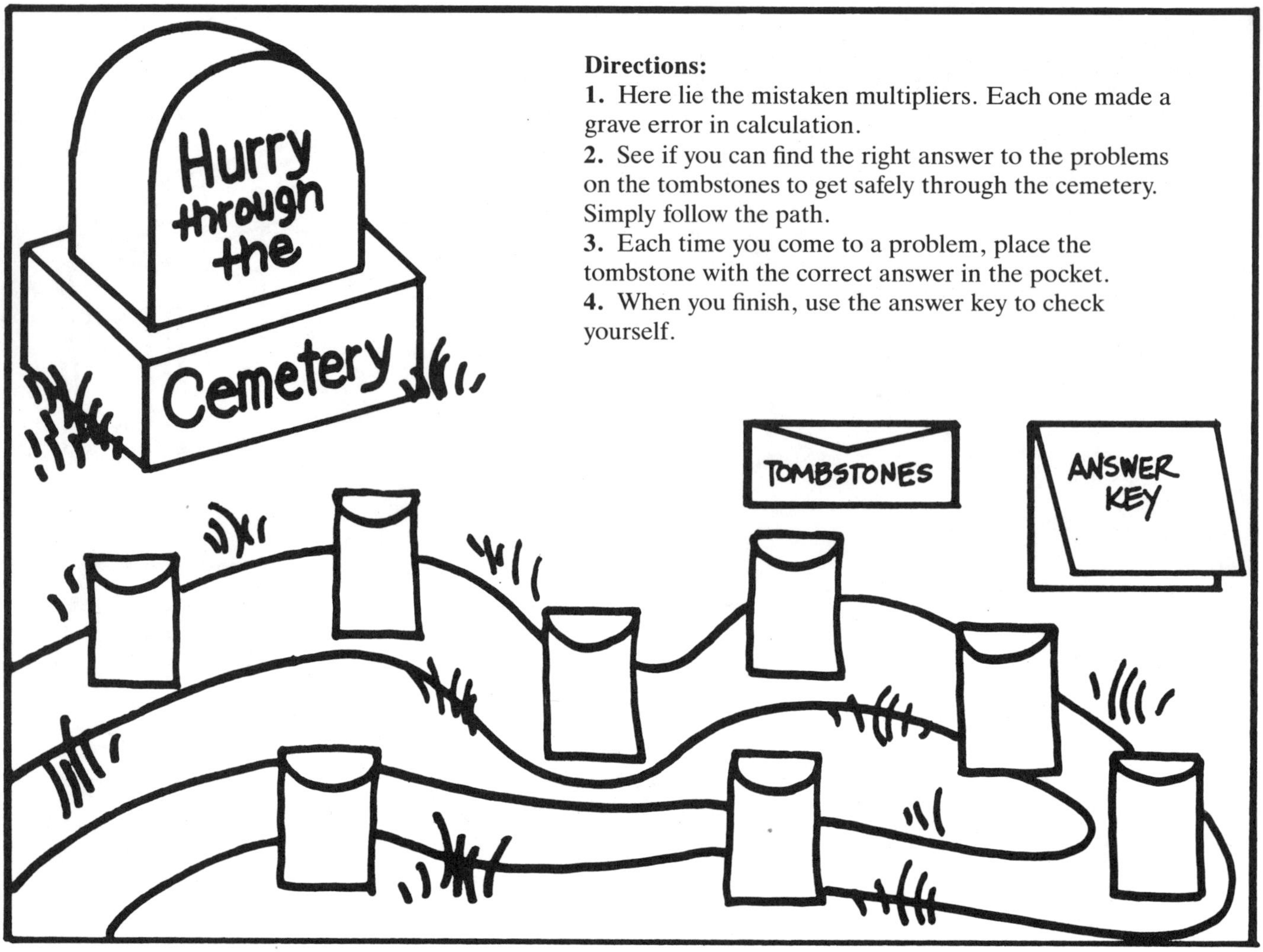

Hurry Through the Cemetery

Skill Reinforced: Multidigit multiplicand and multidigit multiplier

Materials Needed:

- gray background paper
- gray posterboard
- gray construction paper
- felt-tipped pen
- small Jell-O boxes
- 8 library-book card pockets
- scissors
- stapler
- glue

Construction Directions:

1. Use an opaque projector to trace and cut the title tombstone from the gray posterboard. Attach the Jell-O boxes to the board and glue the tombstone to these. This will give a three-dimensional effect.

2. Use an opaque projector to trace the figures and lettering onto the background paper.

3. Copy each of the following problems/answers onto a different library-book card pocket and attach them to the board as shown in the illustration.

541 × 652 = 338,032
103 × 875 = 11,375
296 × 671 = 198,676
156 × 345 = 43,720
356 × 802 = 287,572
303 × 797 = 26,301
406 × 507 = 23,322
177 × 230 = 40,777

4. Cut the gray posterboard using the following tombstone pattern. You will need eight tombstones.

5. Copy each of the following answers onto a different tombstone as shown in the illustration.

352,732	285,512
90,125	241,491
198,616	205,842
53,820	40,710

6. Mark "Tombstones" on the envelope, attach it to the board, and place the tombstones in it.
7. Fold the gray construction paper to form an answer key and mark the front cover as shown in the illustration. Write the following on the inside of the answer key and attach it to the board.

$541 \times 652 = 352{,}732$
$103 \times 875 = 90{,}125$
$296 \times 671 = 198{,}616$
$156 \times 345 = 53{,}820$
$356 \times 802 = 285{,}512$
$303 \times 797 = 241{,}491$
$406 \times 507 = 205{,}842$
$177 \times 230 = 40{,}710$

Learning Board Use: Students take the tombstones out of the envelope and "walk" through the cemetery by finding, on the tombstones, the correct answers for each problem. They place one tombstone in each pocket as they follow the path. When they finish, they may use the answer key to correct themselves.

The Amazing Rainbow Maze

Skill Reinforced: Multidigit multiplicand and single- and double-digit multiplier/Regrouping/Number patterns

Materials Needed:
- background paper in red, blue, yellow, green, and white
- yellow posterboard
- green posterboard
- straight pins
- legal-size envelopes
- scissors
- felt-tipped pen
- paper hole punch

Construction Directions:
1. Mount the white background paper to the entire board. Cut the other colored background paper and attach it in the rainbow fashion shown in the board illustration.
2. Use an opaque projector to trace the figures and lettering onto the background paper.
3. Cut and punch the yellow and green posterboard using the pattern shown here.

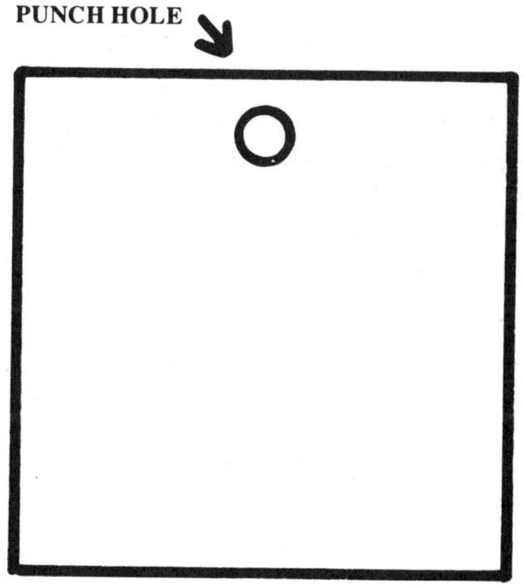

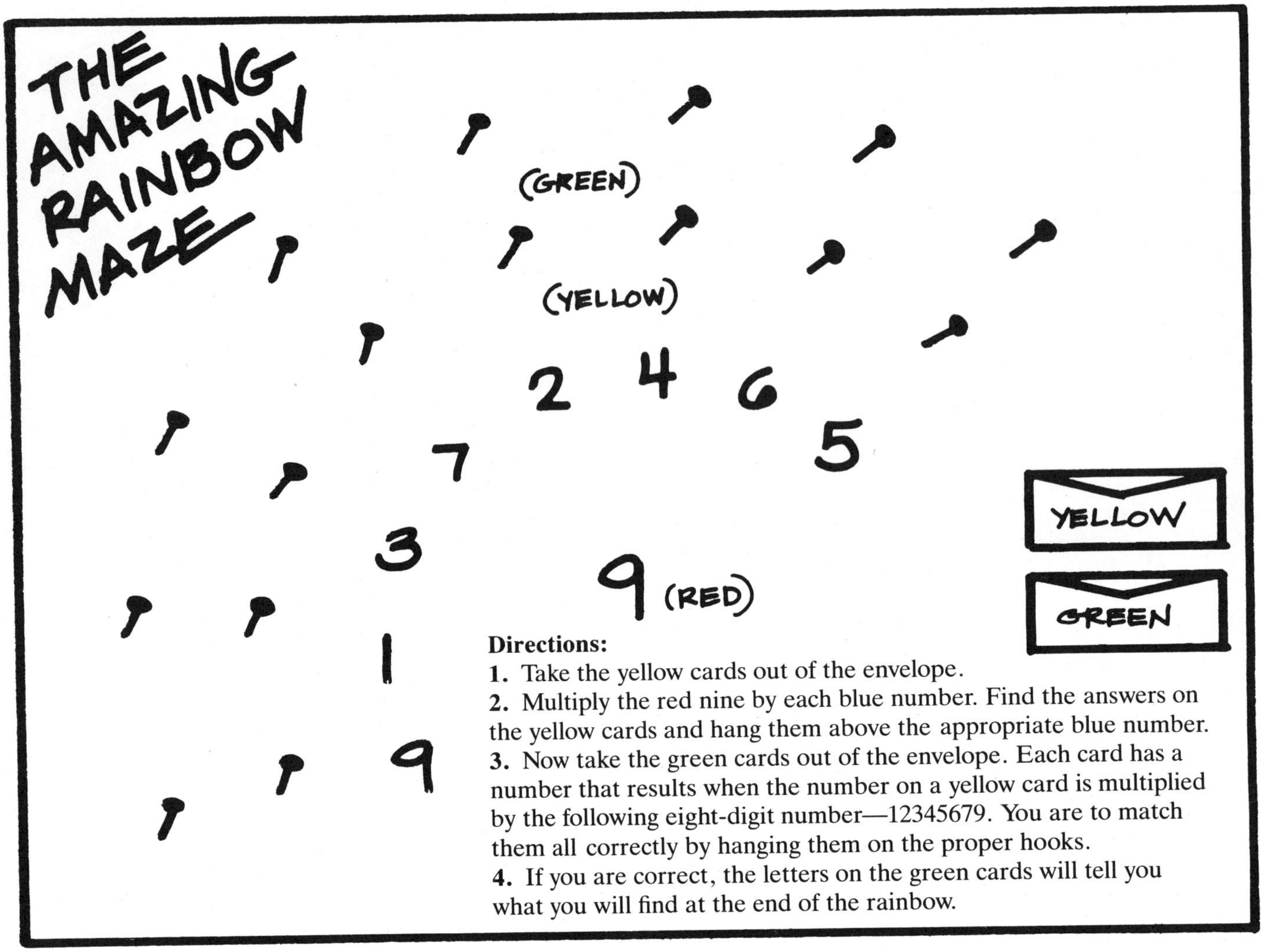

4. Copy each of the following numbers onto a different yellow card.

81, 9, 27, 63, 18, 36, 54, 45

5. Copy each of the following number and letter pairs onto a different green card.

999999999	S
111111111	O
333333333	M
777777777	E
222222222	G
444444444	O
666666666	L
555555555	D

6. Write "Yellow" on one envelope and "Green" on the other, attach both envelopes to the board, and place the yellow and green cards in their respective envelopes.

Learning Board Use: Students take the yellow cards out of the envelope and hang the correct answers on the hooks along the yellow rainbow strip. They do this by multiplying the red number by each blue number. After doing this, they multiply each yellow card by the number 12345679. The answers can be found on the green cards, which are hung on the appropriate hooks. If they are correct, the letters on the green cards will spell SOME GOLD.

PART

V

Special Motivators

Although this entire book is designed to provide motivational activities for students, this section contains some special kinds of motivators. It includes, for example, the idea of private "nooks" and interest centers to which students can retreat to do work. There are also directions for making delightfully original bookmarks, which will encourage a high and constant level of effort, and for making puppets that will attract and hold students' attention.

The versatility of the motivators permits them to be easily adapted for other grade levels, with variations made at the discretion of the teacher.

Airport Mania

Entice your students to do well in math by offering them the following bulletin board activity.

Materials Needed:

- white background paper
- red construction paper
- straight pins
- felt-tipped pens in various colors
- scissors

Construction Directions:

1. Using an opaque projector, copy the illustration of the airport terminal and the directions onto the background paper. Use appropriate colors.

2. Cut out the title letters from the red construction paper and attach them to the bulletin board. You may use the letter patterns provided or make up your own. Attach these to the bulletin board with the straight pins. Pull the letters away from the bulletin board to the heads of the pins. This will give a three-dimensional effect.

AIRPORT MANIA!

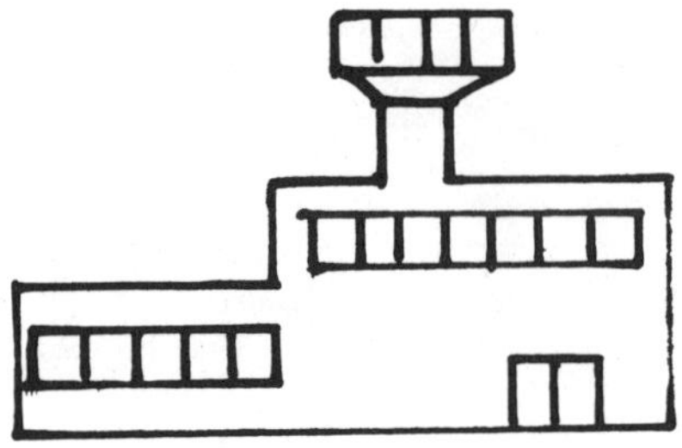

Directions: The next time you receive an A or B on a math paper you may add one object to the airport scene. It can be anything: jets, blimps, birds, a person parachuting down, a witch on a broom, or whatever you wish.

Bulletin Board Use: Every time a student receives an A or B on a math paper (you may want to adjust this scale for some children), he or she may add something to the bulletin board. Just watch the scene expand as the term progresses!

HALF PRICE BOOKS®

Half Price Books
1835 Forms Drive
Carrollton, TX 75006
OFS OrderID 39481376

Thank you for your order, ABE TP Orders!

Thank you for shopping with Half Price Books! Please contact Support@hpb.com. if you have any questions, comments or concerns about your order (738045037)

Visit our stores to sell your books, music, movies games for cash.

SKU	ISBN/UPC	Title & Author/Artist	Shelf ID	Qty	OrderSKU
S465820074	9780673182654	Math Motivators: Puzzles, Games, Bulletin ... Bartch, Marian,Mallett, Jerry J.	EDU3.2	1	

SHIPPED STANDARD TO:
ABE TP Orders
800 Avondale Ave
Columbus OH 43212
help@textbookrush.com

ORDER# **738045037**
Abebooks

Counting Display

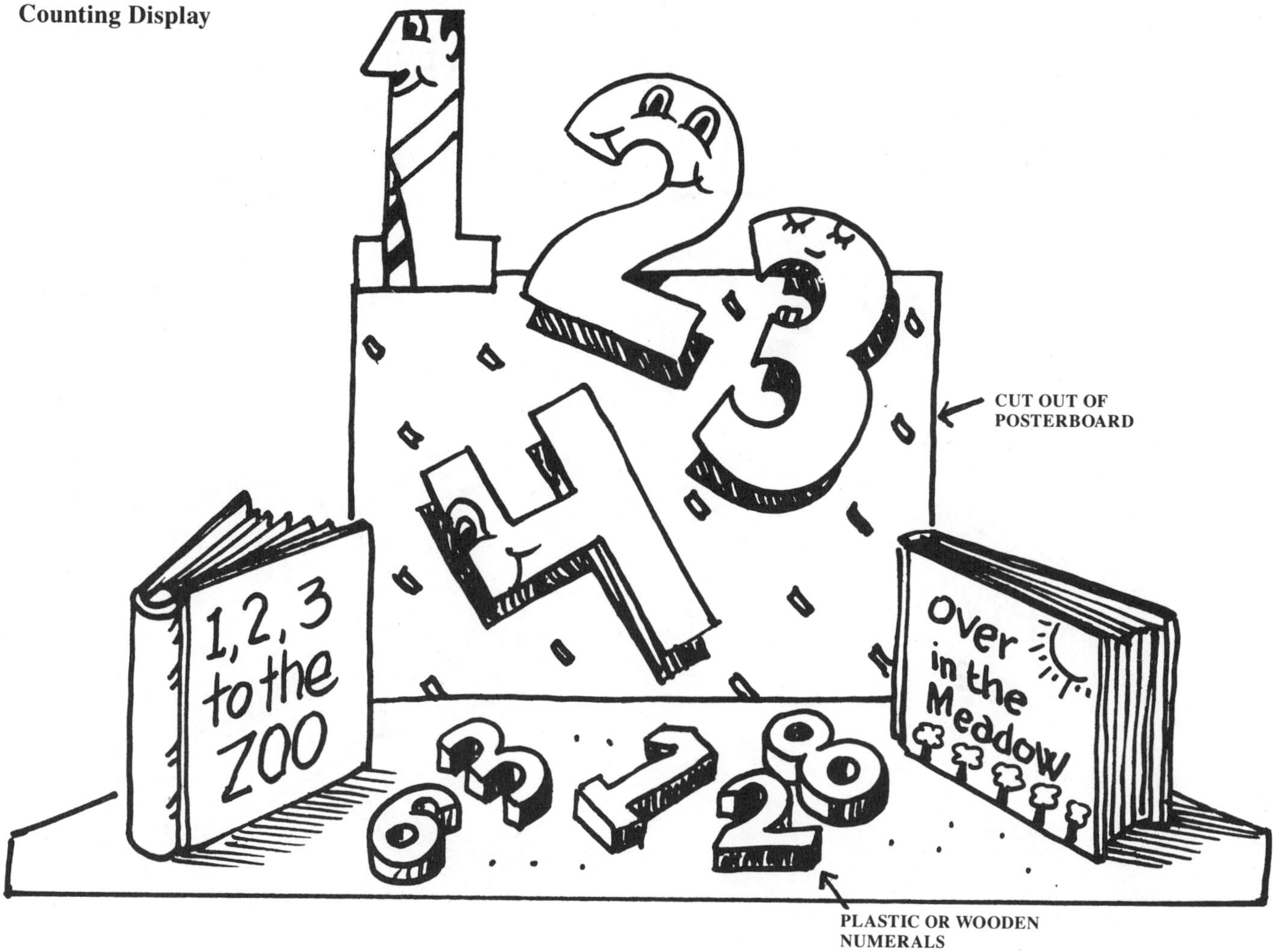

Book Displays

Attractive book displays are a simple way to interest children in reading about math. Try one of the examples shown here and just see if it doesn't spark an interest in math.

CAUTION: Book displays must be changed often to maintain the children's interest.

Books you might want to include in the counting display are:

One, Two, Three, Going to Sea, by Alain
Numbers, by Robert Allen
One, Two, Three, by Marc Brown
1, 2, 3, to the Zoo, by Eric Carle
The Very Hungry Caterpillar, by Eric Carle
Dancing in the Moon, by Fritz Eichenberg
Six Foolish Fishermen, by Benjamin Elkin
Moja Means One, by Muriel Feelings
Count and See, by Tana Hoban
Over in the Meadow, by Ezra Jack Keats
Nine in a Line, by Ann Kirn
3 × 3 Three by Three, by James Kruss
One Dancing Drum, by Gail Kredenser and Stanley Mack
A Spooky Story, by Bill Martin, Jr.
Numbers of Things, by Helen Oxenburg
Numbers, by Jan Pienkowski
Numbers, by John J. Reiss
One Was Johnny (Nutshell Library), by Maurice Sendak
1 Is One, by Tasha Tudor
Brian Wildsmith's 1, 2, 3's, by Brian Wildsmith

Grabbag Display

Books you might want to include in this grabbag display are:

Names, Sets and Numbers, by Jeanne Bendick
The I Hate Mathematics! Book, by Marilyn Burns
Finding Out About Shapes, by Mae Freeman
Math Menagerie, by Robert R. Kadesch
How to Count Like a Martian, by Glory St. John
Games with the Pocket Calculator, By Sivasailam Thiagaragan and Harold D. Stolovitch
Shadow Geometry, by Daphne Harwood Trivett
Funny Number Tricks, by Rose Wyler and Gerald Ames
Nutty Number Riddles, by Rose Wyler and Eva-Lee Baird

A Garden of Math

Use the "garden of math" characters for self-checking activities to reinforce a variety of math skills. These plant characters are all made from posterboard cut into the shapes shown on pages 39 and 40. Cut slits and slide in the white posterboard strips as shown in the illustrations. Then print the problems or equations vertically on the left strip and their corresponding answers on the right one. Be sure to jumble the problems and answers so they do not appear directly across from each other when both strips are passed through the pieces from beginning to end. Now print the problems along with their correct answers on the backside of each piece.

Sunflower Subtraction

Apple Addition

Metric Mushroom

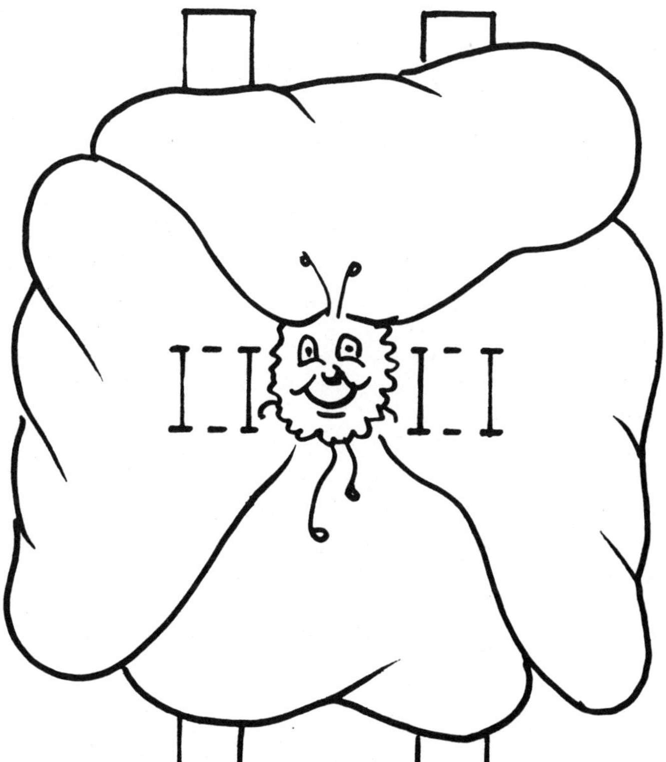

Dividing Daffodil

Multiplying Morning Glory

Fraction Fir

Math Mobiles

Make a math mobile and hang it in your classroom. Simply cut out the different shapes as shown and attach them with string. By using an opaque projector you can easily trace the shapes on the posterboard. Now encourage your students to make their own "math mobiles."

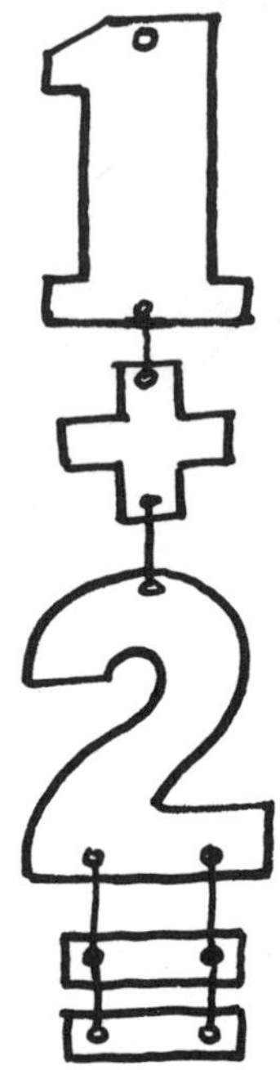

NOTE: You might want to hang these in the classroom and have a contest to see how many students can work all the math-mobile problems.

Calculatin' Castle

As shown in the illustration, a large cardboard box can be transformed easily into the front and sides of a castle. Simply use a razor-blade knife to cut out the door. Then add the medieval effect by illustrating the box with some latex paint. Fold back the sides of the box and use masking tape to attach the sides to the wall.

NOTE: Be sure to draw or paint the windows. Do not cut them out, because this will weaken the structure of the box.

Who's Who in Math

At the conclusion of several six- or nine-week periods, compile a book entitled *Who's Who in Math.* Each child who has a high average, or who has improved dramatically from the previous period, should write a brief autobiography. You may want to use the form on page 54 for this purpose.

Once you have collected all the autobiographies, staple them in a cover similar to the one pictured above, and place it in a conspicuous location in your classroom. Students will enjoy reading about their classmates and the book will provide incentive to have their own personal sheet appear in the next book.

SUGGESTION: You might have a special section in the book for winners of math games, high scorers on certain math sheets, etc. This will help certain students participate whom otherwise might be unable.

Math 50

You should copy the following onto your chalkboard:

MATH 50

Team A — Team B

1–9 people: receive one point each
10 or more: receive two points each
All up: receive 50 points and begin again

Now divide your students into two teams. Choose a team to start and give the first student a problem. If he correctly answers it he stands and you give the next person a problem. This continues until someone on team A misses a problem. When this happens, you total up the score, using the formula on the board, and begin again with team B.

NOTE: If all team members correctly answer their problems, the team receives 50 points and starts again.

Monstrous Math

Divide your students into two teams and have them sit on opposite sides of the room. Give each person on both teams a number beginning with 1. Now copy the following onto the chalkboard. Be sure to number as high as there are persons on each team.

You begin the game by circling one numeral for each team. These do not have to be the same numeral. Team members with corresponding numbers go to the chalkboard. You give them a problem. They must turn to the board and correctly write the answer. When finished, they turn around and face the class. The first one to do this receives a point for her team. But if the first person finished has answered the problem incorrectly, her team loses a point.

The game continues until all the numbers have been circled. The team with the most points wins.

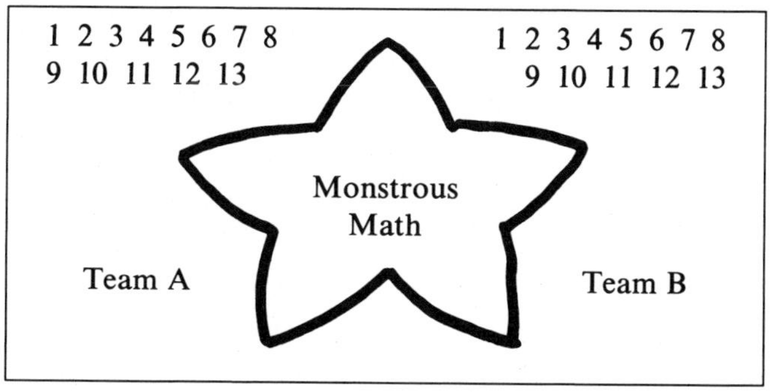

Beat Mr. Dumas

Divide your students into two teams and have them sit on opposite sides of the room. Now copy the following onto the chalkboard:

Start	
B E A T D R.	D U M A S
	Safety Zone

Choose a captain for each team and have them stand next to the word "Start." Call on a student from team A to answer a problem. If correctly done, the captain moves to the letter "B" on the board. Now do the same with a member from team B. Continue alternating between teams, calling on different team members to answer problems. At each correct response, the representative captain moves to the next letter on the board. If a team member incorrectly answers a problem, the captain must return to "Start" and begin over unless that captain has advanced into the Safety Zone. In this case, the captain only has to return to the safety zone line. The first team to move their captain to the last letter is the winner.

Capture the General

Divide your students into two teams and have them stand on opposite sides of the room. Place a book on each of four desks located on each side of the room. Now the two teams turn around and you place a small sheet of paper with "General" written on it under the cover of one of the books on each side of the room.

Call on the first student on team A and say, "Where do you think the general is?" Whichever desk she says—desk one, two, three, or four—that is the number of problems she must correctly answer. For example, if she said, "Desk two," she must answer two problems correctly in order to look for the general in that book.

Whenever a team member "captures a general," a point is added to that team's score and you hide the general again. The team with the highest score at the end of the game is the winner.

Worksheets

THE BIG CHALLENGE

Match the number sentences with the list of properties. As you do so, write the circled letters in the properties in the space next to the corresponding number sentence. If you are correct, these letters will spell a word that tells you how well you did!

____
Name

____
Date

_____ $(6 + 4) + 3 = 6 + (4 + 3)$

_____ $5 \times 6 = 6 \times 5$

_____ $1 \times 8 = 8$

_____ $9 + 0 = 9$

_____ $3\,(9 + 8) = (3 \times 9) + (3 \times 8)$

_____ $(7 \times 2) \times 8 = 7 \times (2 \times 8)$

_____ $-6 + 6 = 0$

_____ $2 + 3 = 3 + 2$

_____ $8 \times 1/8 = 1$

List of Properties

1. distributive-multiplication ove(r) addition
2. as(s)ociative-multiplication
3. inverse-addi(t)ion
4. comm(u)tative-multiplication
5. id(e)ntity-addition
6. a(s)ociative-addition
7. identity-multi(p)lication
8. commutative-(a)ddition
9. inve(r)se-multiplication

DOWN KILLER HILL!

Name ______________________

Date ______________________

Take a ride down Killer Hill! Each time you come to a problem, look for the answer in the snowman. You are to mark the correct answer and letter next to the problem in the space provided. If you are correct, the letters will spell how long it took you to coast down the hill.

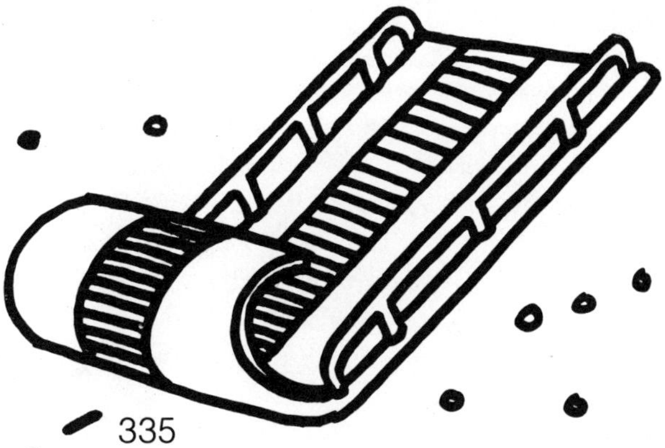

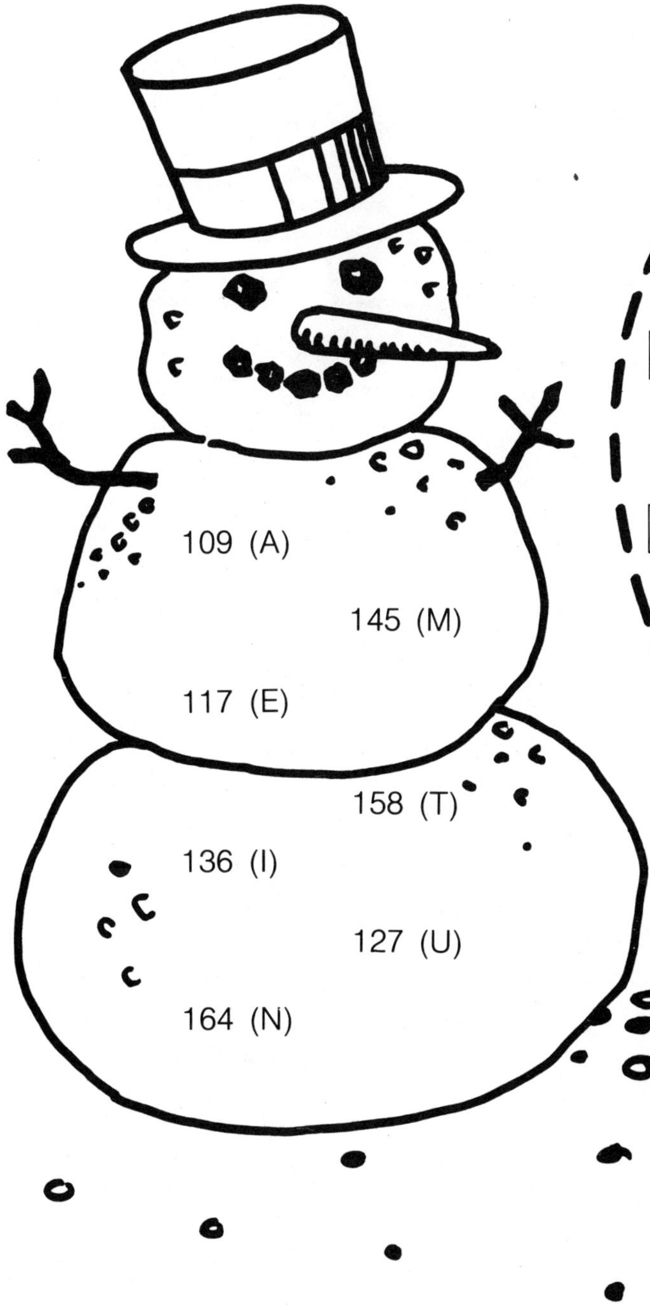

335
−226
☐ ____

672
−527
☐ ____

552
−416
☐ ____

783
−619
☐ ____

646
−519
☐ ____

487
−329
☐ ____

994
−877
☐ ____

TRIP THROUGH SPACE

Name ______________________

Date ______________________

Begin at the space ship and follow its path. Each time you come to a star look at the problem inside it. Can you find the answer in the planet? Copy down the correct answers along with their letters in the spaces provided. If you are correct, these letters will spell where the space ship is traveling.

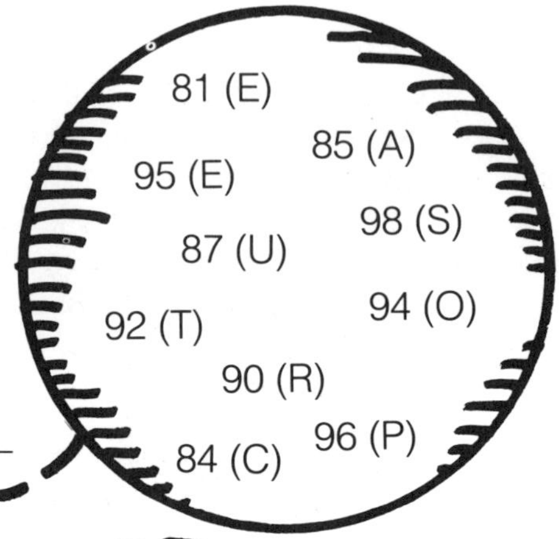

$2 \times 47 =$ ______

$3 \times 29 =$ ______

$4 \times 23 =$ ______

$5 \times 19 =$ ______

$6 \times 15 =$ ______

$7 \times 14 =$ ______

$8 \times 12 =$ ______

$5 \times 17 =$ ______

$6 \times 14 =$ ______

$3 \times 27 =$ ______

YOU'LL NEVER GUESS

Name ______________________

Date ______________________

"Please tell me," begged Sue.

"No. It's a secret," said Bobby.

"Well, can I guess?" asked Sue.

"You can if you want to, but you'll never, NEVER guess it," answered the boy.

Sue and Bobby were riding home from school on "Old number five." This is what they called their school bus. Bobby had told his good friend Sue that he and his dad had gone shopping and gotten something for Sue's birthday. But now she wanted to know what it was.

"Is it a game?" asked Sue.

"You'll find out tomorrow on your birthday," said Bobby.

"Oh! This isn't fair!" cried Sue.

Why don't you help Sue? It's easy to find out what Bobby is going to give her for her birthday. All you have to do is solve the puzzle on this page.

Look at the row of problems. Now look at the wrapped present. Inside are the answers to these problems. You are to copy down the correct answers along with their letters in the space next to each problem. If you are correct, the letters will spell what Bobby bought Sue for her birthday.

$10 \div 2 =$ ____________

$18 \div 6 =$ ____________

$16 \div 4 =$ ____________

$11 \div 1 =$ ____________

$22 \div 11 =$ ____________

$40 \div 5 =$ ____________

$56 \div 8 =$ ____________

$42 \div 7 =$ ____________

$72 \div 8 =$ ____________

$39 \div 3 =$ ____________

$28 \div 2 =$ ____________

$100 \div 10 =$ ____________

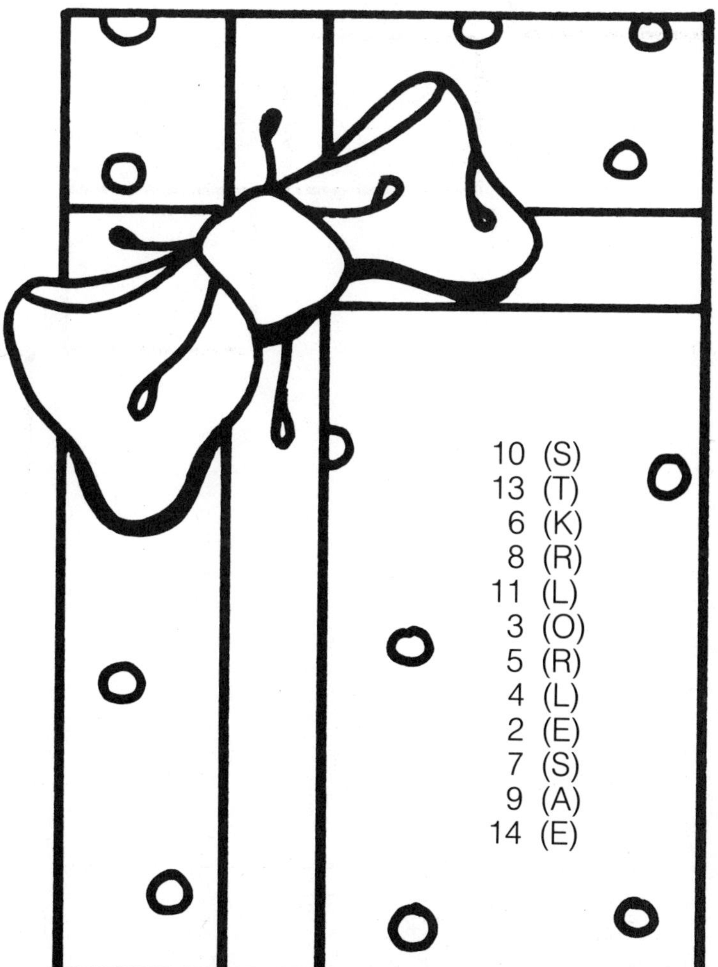

WANDA'S ROCKET

Wanda seems to be having rocket trouble. It looks like she might be in for a CRASH landing! How do you think she feels right now? There is a way for you to find out. Cut out the answer cards and place them under the correct problems in Wanda's path. Be sure to work the problems in base 5. If you are correct, the letters on the answer cards will spell how Wanda feels.

41
+32

131
+424

34
324
+311

224
30
+134

23
+44

123
234
+146

1224 A	1003 D	122 E	123 S	1110 C	443 R

RUNAWAY ROBOT

Name ______________________

Date ______________________

Can you guess the name of our "runaway robot"? There is a way for you to find out. Follow the robot's path. When you come to a problem with a correct answer, copy down the letter under it on one of the spaces at the bottom of this page. If you are right, these letters will spell the robot's name.

$-5 - 12 = -17$
(O)

$-12 - 10 = 2$
(B)

$4 - 6 = -2$
(R)

$-3 - (-8) = 5$
(N)

$8 - 2 = -10$
(T)

$7 - (-2) = 9$
(A)

$8 - 8 = 0$
(L)

$-2 - 5 = -7$
(D)

$-4 - 3 = -1$
(E)

$6 - 6 = -12$
(D)

$-10 - 12 = -22$
(M)

$-13 - 30 = -43$
(E)

$6 - (-10) = 4$
(I)

$-4 - 4 = -8$
(C)

$-8 - 2 = -10$
(M)

$6 - (-1) = 7$
(T)

$-7 - (-7) = -14$
(R)

$5 - (-8) = 13$
(A)

$10 - 20 = -30$
(I)

$12 - (-4) = 16$
(L)

DIVIDING ANIMALS!

Name

Date

Below are three "dividing animals" puzzles. You are to look through the numeral list in each cage and decide which ones are divisible by the puzzle's number. You are to copy down the letters next to each divisible number in the spaces provided in each puzzle. If you are correct, the letters in each puzzle will spell a different kind of animal.

Puzzle 1:
Must be divisible by 5.

1981 (B)
1640 (C)
8925 (A)
2463 (I)
2310 (M)
1585 (E)
7682 (T)
9898 (U)
1150 (L)

Puzzle 2:
Must be divisible by 4.

8291 (P)
3432 (H)
5924 (O)
7503 (I)
1811 (J)
1976 (R)
2988 (S)
3069 (L)
6544 (E)

Puzzle 3:
Must be divisible by 3.

4073 (E)
3276 (M)
9060 (O)
1111 (G)
1623 (N)
8117 (P)
7128 (K)
2333 (H)
4197 (E)
1101 (Y)

CROSS OVER THE BRIDGE

Name ______________________________

Date ______________________________

Beginning at the car, try crossing over the bridge. Each time you come to a problem, find the correct answer in the ship. Copy down the answer and letter in the spaces next to the problems. If you are correct, the letters will spell the name of the bridge.

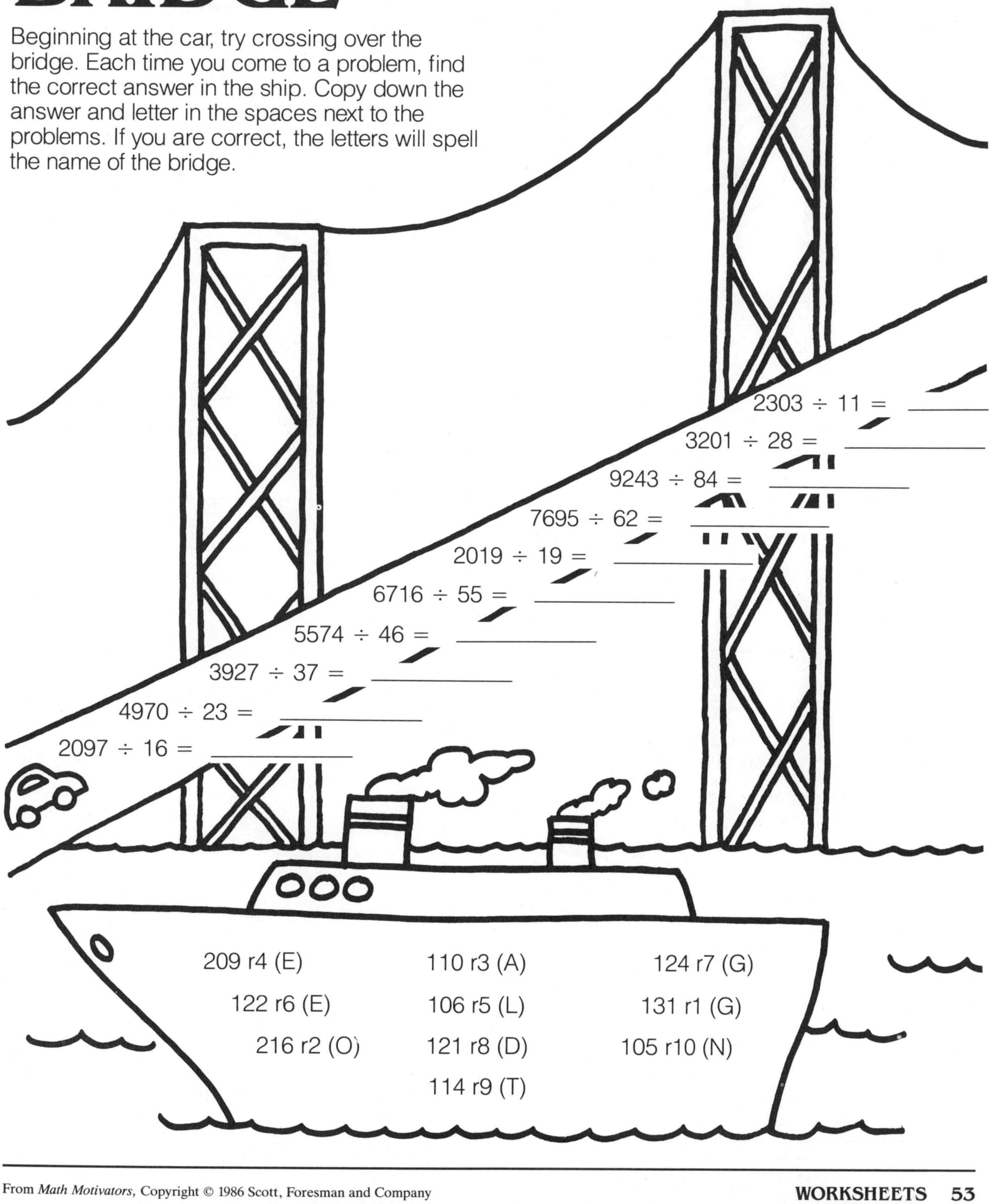

2303 ÷ 11 = __________

3201 ÷ 28 = __________

9243 ÷ 84 = __________

7695 ÷ 62 = __________

2019 ÷ 19 = __________

6716 ÷ 55 = __________

5574 ÷ 46 = __________

3927 ÷ 37 = __________

4970 ÷ 23 = __________

2097 ÷ 16 = __________

209 r4 (E) | 110 r3 (A) | 124 r7 (G)
122 r6 (E) | 106 r5 (L) | 131 r1 (G)
216 r2 (O) | 121 r8 (D) | 105 r10 (N)
114 r9 (T)

WHO'S WHO IN MATH

Name ______________________

Date ______________________

Here is my story:

____

____

____

____

____

____

____

____

____

Here is a picture of what I want to do when I grow up.

This is a picture of what I

____